# From Crisis to Calm:

## A Family Guide to Handling Dementia Behavior

Deborah Bier, PhD

From Crisis To Calm: A Family Guide to Handling Dementia Behavior
by Deborah Bier, PhD

Published by Decoding Dementia, Inc., P.O. Box 411 Lancaster, MA 01523

www.FromCrisisToCalm.com



Cover: Brooklyn Billmaier

**ISBN:** 9781791777470
**Imprint:** Independently published



Neither the author nor the publisher assumes any responsibility for errors, omissions, or contrary interpretations of the subject matter herein. Any perceived slight of any individual or organization is purely unintentional. The names and other identifying information of clients and their families have been changed to preserve their privacy.

The information in this book is for educational purposes only. This information is not intended to be diagnostic or to serve as a substitute for actual medical advice. If you are concerned about your or someone else's health, please seek help from a licensed medical professional.

# DEDICATION

For Rich, without whom life
just wouldn't matter so much to me.

# CONTENTS


| | | |
|---|---|---|
| 1 | There is Hope | 1 |
| 2 | Who I Am | 7 |
| 3 | Some Dementia Basics | 11 |
| 4 | Introduction to The Four Dementia Keys | 25 |
| 5 | Dementia Key #1: Nurture Emotions | 31 |
| 6 | Dementia Key #2: Customize Care | 53 |
| 7 | Dementia Key #3: Communicate Well | 63 |
| 8 | Dementia Key #4: Optimize Stimulation | 75 |
| 9 | Addressing Problems That Drive You Nuts | 87 |
| 10 | Caring for the Caregiver | 97 |
| 11 | Solutions for Your Extended Dementia Crisis | 113 |
| 12 | Conclusion | 125 |
| | About the Author | 129 |
| | Thank You | 131 |
| | Acknowledgements | 133 |

# *IMPORTANT: PLEASE READ AND REMEMBER!*

## IN CASE OF EMERGENCY DO NOT CONSULT THIS GUIDE

*From Crisis to Calm* is a guide for study only. If your family member needs immediate medical attention at home, call 911 or your local emergency number.

If your family member needs immediate attention while in the hospital or other care setting: speak with a nurse, doctor, or other provider on duty.

If your family member is being abused or neglected, call 911, your local emergency number, or adult protective services.

SOME ISSUES REQUIRING IMMEDIATE MEDICAL or LAW ENFORCEMENT ATTENTION MAY INCLUDE:

- Your family member is in danger of self-harm or harming you.
- Your family member's environment is dangerous.
- Your family member's dementia symptoms seem suddenly and dramatically worse: confusion, hallucinations, extreme fear, unresponsiveness, etc.
- You or someone else feel like or are hurting or neglecting your family member with dementia.
- You feel like you are in danger from self-harm or from your family member with dementia.

# 1. THERE IS HOPE

*Al's family was out of patience and completely exhausted. He had Alzheimer's disease, and was so agitated he'd pace around and around inside the house. By sundown, he'd get so worked up that he'd be screaming and crying out. He'd start throwing, ripping, and smashing things within his arm's reach. It was pandemonium, and Al's family were at their wit's end. They tried everything they could think of, and they were out of ideas. It had been weeks of this behavior, and they couldn't go on like this.*

*For a few hours of respite care to give them a break, the family hired a caregiver whom I had trained in dementia behavior handling best practices. As they watched from afar, the family saw Al's agitation diminish quickly as she cared for him. By a couple of hours, Al and the caregiver were chatting and laughing together, having a wonderful time. The destructiveness, agitation and screaming were gone and didn't reappear the rest of the time the caregiver was present. The family was simply astonished!*

How did the caregiver do it? The answer is this book.

So, thank goodness you're here… *and not a moment too soon.* Your family is in dementia behavior trouble, and it can feel like your sanity is hanging by just a slender thread.

You're caring for your family member with Alzheimer's, or another type of dementia, and you're naturally become frustrated, angry, sad, and exhausted beyond words. You've come to the right place.

Perhaps you're lonely, isolated, and anxious because dementia behavior has swallowed your life, your family, friends, and job, and you're fantasizing about walking away and disappearing forever. I'm so glad that you're arrived here.

You just can't stand it anymore and are ready to put your family member – perhaps your mother? – into a memory facility. But you're wracked with guilt and don't know what to do. You'll find help here.

Maybe your mother with dementia is fighting and acting out all the time. You've arrived at the very right place...and without a moment to spare, I'm guessing. I'm so glad because we should be able to calm many of her behavior troubles.

As you work your way through this book, you'll find simple-but-powerful, easy-to-learn methods to not only calm these problems, but *prevent* many of them from happening again in the future. I'll teach you methods I've taught to *thousands* of family caregivers – methods that ended their crisis, and stopped or greatly reduced ongoing struggles and fights around daily care. (You'll also see in Chapter 3 why I am guessing you are female and that your family member with dementia is your mother.)

I want to give you hope that your life as a dementia caregiver can change for the better. If you learn and *consistently* use the care solutions in this book, the future can be better. Not just for you and your family member with dementia, but for *everyone in your family who also learns and uses these methods regularly.*

Using these care methods myself, I've sometimes seen significant improvement right away, literally in seconds. When you're using the absolutely right approach, the turn-around can be very rapid, even immediate. Because everyone is different, and because you won't always know *"the" absolutely right intervention* at *"the" absolutely right moment*, sometimes the turn-around can take longer.

You may need to engage in some trial and error to nail-down the right changes. And sometimes, the real problem isn't you or your mother but someone else stirring her up. But with the help of this book, you *can* make improvements. You don't have to live in this turmoil the rest of your mother's life. We can calm this dementia crisis…together.

That this disaster could *ever* turn around, much less possibly quickly improve, might seem impossible to imagine, yes? Of course it is; that's only natural. I accept that what I'm saying may be a surprise for you and your family. But I've seen it happen again and again, and it could happen for *you*, as well.

I want to give you hope here because feeling *hopeless* often leads to feeling *helpless*, and those are such painful feelings to carrying around all the time. Helplessness and hopelessness keep all of us from seeking solutions, or even believing solutions might exist. At this moment, it's important to have hope because it will give you the energy you need to learn and practice The Four Dementia Keys in this book.

So, how are we going to work this miracle? Well, no magic wands are actually needed, although the positive change you bring about can at times seem nothing less than magic! Using The Four Dementia Keys in this book, your family member could:

- Be more calm and peaceful

- Enjoy life more

- Have an improved relationship with you and others

- Work with you and accept daily care

- Have fun

- Function better
- Use less of your time and energy for daily care
- Stop taking over your life and making you crazy

In turn, your mother with dementia might become more recognizable to you. You might be able to better resume some parts of your prior relationship, though it will be different than before dementia. That person you know and love is still "there", even though you haven't seen her perhaps for weeks, or even months. I'll teach you how to calm your crisis so that your mom could be at least *sometimes* more herself again, as long as she's not yet entered the very end stage of dementia.

The truth is that people with dementia require special care methods, and no one's born knowing how to provide all of these correctly. They must be *learned* to get good results. Some family caregivers intuitively figure out a couple of improved approaches, though they probably can't tell you how or why what they're doing works. They also likely can't teach these methods to others, because they themselves don't necessarily know what their magic touch consists of. In this book, I'll teach you just enough information about dementia so that you know why as well as how. You'll need this knowledge because you will be tweaking The Four Dementia Keys to work best with your family member, so you'll need to understand some of the operating principles behind these Keys.

Did you know: no one in the medical system has the job of teaching you and your other family caregivers the exact solutions you need to get out of the care nightmare you're in, much less how to have prevented the crisis in the first place? Not many people working in the medical system today know these behavior interventions exist, and those few certainly don't have the time to teach them to you. I also wasn't taught about these solutions in my formal training as a psychotherapist and counselor… dementia wasn't even mentioned except in passing and never a single word of what I'll teach you here.

This is frustrating because it's recognized with other diseases, such as diabetes, that patient and family education are truly key to getting the good outcomes we all desire. For example: there are nurses with special training who hold the title of Diabetes Educator, and sessions with them are generally covered by insurance. But dementia behavior? No – no one in mainstream medical care is teaching the methods you'll learn here. And that's a heartbreaking pity to me…luckily, one I can help address with this book as well as the further trainings and consultations I offer families.

In fact, likely much of what you'll learn in this book is very different from you've been doing, and how most people doing behavior operate. Some parts of the Four Dementia Keys will seem counter-intuitive to you. Others will be *exactly* the opposite of what we do when caring for people *without* dementia. In the process, you'll quickly see why things have gotten out of hand. With every chapter, you'll know more how to get back on track. You can learn to put out the fires of dementia behavior quickly with what you'll read here.

Research shows us that improving behavior requires learning not just new methods, but for the people learning them to receive a period of ongoing mentoring and coaching as part of the process. This is why this book is structured to be worked with over six weeks. The practice you'll get over that time will allow you to nurture your new skills. Quickly, they'll make sense to you, and you'll develop an ease and familiarity using The Four Dementia Keys that are now just coming onto your radar.

Everyone stumbles around a bit when learning new skills and we *all* need ongoing mentoring and coaching… even me! This is especially true when working with members of my own family who have some type of cognitive impairment. I make mistakes dealing with my family members that I would *never* make if I were instead dealing with yours. Caring for your own family is very different from caring for a stranger's family member. With our own, we have long-practiced roles and relationships we play, ongoing emotional needs, and old relationship baggage that surfaces when we care for them, and this clouds our view in the moment. But once I step away, I quickly can see what I've done incorrectly and what would have been the better approach.

This will happen with you, too, and it might make you feel stupid. I assure you that making and recognizing errors is an important part of the learning process, and one you can't escape. Please don't beat up on yourself *when* (not *if*) this happens. I will remind you of this again as you go through this book.

What's also going to happen as you work your way through this book is that you'll quickly notice that you're *still* not using all the right moves *all* the time, and you may feel terribly guilty and upset with yourself. I'm bringing this to your attention now because if this happens, you must remind yourself of the following, which I believe to be utterly true:

**I have been doing my very best.**

**Now that I know better, I will do better.**

**New skills take time to develop until they become automatic.**

**I forgive myself, and let go of any guilt or anger I have.**

**I forgive myself for not being a perfect caregiver.**

The methods you'll learn in this book work with all types of dementia, especially Alzheimer's, and to a large extent vascular, Lewy Body, and frontotemporal dementia, and other types that are more rare. When working with people with cognitive impairment due to other causes such as brain injury, or cancer in or of the brain, these solutions can also help to quickly put out the fires of a care crisis. As a result, you'll end up with greater ease as you cross paths with folks with such a history.

Coming up in Chapter 3 is some information about dementia that I want to make sure you know because it will help you understand The Four Dementia Keys you'll learn to use in the following chapters. I'll then show you how The Four Dementia Keys can be put into action when dealing with some of the most frustrating and infuriating behavior issues – perhaps the very ones that are driving you nuts right now. Chapter 10 focuses on you as the engine that pulls the whole behavior train. Chapter 11 offers resources outside of this book, and Chapter 12 is a brief review and an invitation to work further with me.

Though this program is formatted to be done over six weeks, you don't have to take exactly that amount of time. You can do it faster or slower. You might want to take longer to work with Chapter 5 because Dementia Key #1, Nurture Emotions, is the most important of all the Keys. You might digest and use the information here in fewer than six weeks…or take longer if you need to. However it goes, you should do it at your own speed. So, when you see the recommended schedule below, you decide how you want to work with this material. I suggest:

Chapters 1, 2, and 3 during Week 1

Chapters 4 and 5 during Week 2

Chapters 6 and 7 during Week 3

Chapters 8 and 9 during Week 4

Chapter 10 during Week 5

Chapter 11, 12 during Week 6

There's also an online From Crisis to Calm Tool Kit that you should download to accompany this book. There are several items in it to help you use the information here more effectively. Go to www.FromCrisisToCalm.com and click on the "Tool Kit" button.

Throughout this book, I've used examples from many clients' and students' cases. However, to ensure their dignity, privacy and confidentiality, I've changed their names and other identifying features. As I've done with Al's story at the start of this chapter.

# 2. WHO I AM

I want you to know who you've come to for help.

I have a master's in counseling psychology and a doctorate in therapeutic counseling. I have decades of experience working with medically complicated patients, as well as serving as an educator and director of care for home care agencies. I have certifications in gerontology, and dementia behavior and education.

I want you to know that *none* of the above made me understand dementia the way I do now, nor did it give me the hope I can offer you.

My understanding began in 1992 when a car accident left me with a traumatic brain injury. During the next seven years, while trying to meet the obligations of work and home, I struggled daily with vertigo, nausea, insomnia, and an inability to recall or say many words. When reading, by the time I reached the end of a sentence I couldn't remember what preceded it. And although I'd cooked professionally and enjoyed it as a pastime, I'd forgotten how to cook or even what my favorite dishes were. I memorized my own phone number only after a year of practice.

My moods were unpredictable, often extreme, all over the map. I developed a crippling fear of leaving the house for any reason. And when I did manage to leave, I couldn't tell if the faces I encountered were those of agreeable acquaintances, people I didn't particularly like, or of total strangers I'd once met in passing. The one-mile drive from my home to the drug store engendered repeated panic that I didn't know where I was headed because I would repeatedly forget – even in such a short amount of time. Unsurprisingly, my anxiety often overwhelmed me, my erratic emotions taxed my husband's patience, and my depression was crushing.

I want you to know who I am because you will probably recognize that my symptoms, and my experience of my symptoms, much resembles what we see in a dementia like Alzheimer's.

By recognizing in my experience what you already know and have seen, you know that I understand the experience of dementia from an unusual double-pronged view: one from inside cognitive impairment, and another from the point of view of the caregiver. These reach far beyond my formal credentials.

I and my unusually loyal husband had to find our own way on this journey because neither of us understood how my symptoms and behavior were related to my injuries.

But even during the worst of my brain damage, I was able to create workarounds for some of my symptoms. When venturing cautiously outside, afraid I'd come across someone who felt familiar but whom I was unable to name, or place in any kind memory-jogging context, I eventually learned that if I took a moment to push aside my anxiety and concentrate on the person's face, I could trust how I felt about them. In time, I trusted that I felt good about people I liked. I felt bad when I recognized people but felt dislike for them. And I felt nothing in particular when I encountered total strangers. I also learned that I could manage simple errands better by taping a note to my steering wheel that read, "Go to the bank and then to the pharmacy."

Time passed. I pieced together many more practical methods for working around my brain injury. Now long after my car accident, I still have some memory issues, difficulty with numbers, and can sometimes be rattled by everyday tasks that are a breeze for other people. But I now function quite well in most areas of my life.

As I made the transition to full-time work in home care, learning more about behavior from colleagues, researchers, and dementia clients, I interpreted their approaches and advice through the lens of my living for years with cognitive impairment. What's been fascinating to learn is that the strategies and methods I developed for overcoming my dementia-like difficulties are actually best practice in behavior. My life is the foundation of my work.

You now know who I am and the reasons why I am optimistic about the hope I can give you.

In the following pages, I will bring my different perspectives, professional knowledge, and personal insights to providing you with practical evidence and experience-based methods and strategies for stopping your dementia behavior trouble in their tracks.

**Fast Summary – Chapter 2**

- People with cognitive disabilities from a brain injury can be similar to people living with dementia

- Dementia care skills are not included in training of healthcare professionals

- Experience is a vital teacher in providing excellent behavior

# 3. SOME DEMENTIA BASICS

You're exhausted from providing 24-hour dementia behavior handling, and you have many questions that you need to have answered. *But not all information about dementia is going to help you be better at behavior or more effectively meet your current crisis.*

You could easily drown in information that's not essential to resolving your caregiving problems, so I'm not going to discuss neuroanatomy at length, nor am I going to use and explain a lot of medical terminology.

You're already overwhelmed and had more than a lifetime's worth of frustration, so I'm not going to walk you through the complexities of dementia's causes, dementia staging, or dementia medications.

In this chapter, my primary goal is to give you the amount and kind of information you need to speak clearly and effectively in language that both you and your family truly need.

Why is it so important *right now* for you to gain confidence in this knowledge and how you use and talk about it? If you can't understand the true nature of your family member's dementia, it's extremely hard for you to know what types of disabilities she's dealing with.

If you don't know what types of disabilities your family member is dealing with, you'll have a difficult time distinguishing which of her *abilities* are still fairly intact.

If you can't accurately determine your family member's intact abilities, you won't be able help her exercise them, and she'll more quickly lose her ability to do what she still can do. *Use it or lose it* is the rule that's central to effective dementia behavior handling.

I suppose you could *guess* about your family member's losses and strengths, but guessing will cause you to waste time and make mistakes. Right now, I want to introduce you and your family to ten vital areas you need to learn about *before you begin to put into practice the methods and advice I'm going to cover.*

Yes, ten vital areas that you *and your family* need to learn.

Unless you live *alone* with your mother and are her *only* caregiver, I strongly recommend that you make a consistent effort to share what you learn from me with *everyone* who's regularly around your mom. By letting others know about the solutions I offer, you'll avoid working with your mother at cross purposes.

When *everyone* who's in your mother's life shares the same information, intentions, and goals, the progress you make in beautifully fine-tuning your mother's care won't be so easily monkey-wrenched by uninformed errors of judgment and hair-raising inconsistencies.

Trust me on this. If you don't let others know what you're learning from this book, you run the risk of bringing more crisis than calm to your dementia behavior issues.

## Dementia is Not Normal Aging

Although it's commonly believed that dementia is what you get for growing old, we all know very old people whose minds are still sharp and clear. This is because dementia is a *disease*, not a consequence of normal aging. Dementia is very different from the slowing of memory-recall that many people experience as they age.

## Dementia and Alzheimer's Are Not Interchangeable Terms

These words are so commonly confused that I need to make sure you know what they mean so that you'll understand what I'm talking about. By having a good grasp of the difference between dementia and Alzheimer's, you and your family will better understand your mother who is living with dementia.

Dementia is an umbrella term for dozens of diseases that cause cognitive impairment. Alzheimer's is one of those many diseases. It's the one we hear about the most because Alzheimer's comprises 60% to 80% of all dementia diagnoses.

However, in the same way that baseball is a sport, but not all sports are baseball. All Alzheimer's is dementia, but not all dementia is Alzheimer's. You'll likely hear or read about types of dementia other than Alzheimer's. The most widespread are:

- Mild Cognitive Impairment
- Lewy Body Dementia
- Vascular Dementia
- Frontotemporal Dementia

Each type of dementia has its own unique set of symptoms. Each type of dementia progresses along its own unique path. For example, although individual cases vary widely, I can anticipate certain things in someone with frontotemporal dementia that I won't necessarily expect in someone with Alzheimer's disease.

## Dementia Is NOT the Same as Memory Loss

You might think that dementia involves only memory loss. While I'm not sure why or how it started, the routine association of these terms reinforces the misunderstanding: *memory units* are residential homes for people with *dementia*, and *dementia* is said to be a *memory disease* that you can have diagnosed at a *memory clinic.* Cementing dementia and memory loss together isn't only highly misleading and incorrect, *it's actually preventing you from gaining a useful understanding of your situation.*

Your mother's dementia is about much, *much* more than memory loss.

Your mother's dementia is also about a decline in her *ability to function* and the *loss of abilities* she once had. This decline and loss are typically accompanied by significant personality changes.

As true as it is that people with Alzheimer's have a significant *memory* disorder, it's equally true that they also have symptoms related to disorders of *movement, thinking, immune function, mood, self-control, and language.*

Because dementia damages the brain across so many domains, and because some types of dementia don't involve memory-loss symptoms, it's far more accurate and useful to refer to dementia as *brain failure.*

Frontotemporal dementia, for example, has three different variants: behavior, language, and movement. Please note that, especially before their advanced stages, these variants *don't include prominent memory loss.* However, memory loss is one Alzheimer's early symptoms, and it continues throughout the course of the illness. Each specific type of dementia is talking about a different set of symptoms and abilities preserved.

What's of special interest is that people with dementia don't lose *all* types of memory. For example, in most people with dementia, their memory of poetry, music, and lyrics usually remains intact until the end. Their long-term memory from about ages 5-25 is retained for a long time. Even when they've otherwise lost the ability to understand or use speech, they can often *perfectly* sing the words of an old favorite song.

**Dementia Harms the Brain in Ways That Can't Be Seen**

Dementia takes place on a cellular level, hidden from our view. What we can observe directly are brain failure's outward manifestations. This is why family members often struggle with understanding that their loved one with dementia has a physical disease. They end up saying things like, "*If only Mom tried harder*, I know she'd remember," or "*If only I explained the facts* to Dad in a different way, or again and again, he'd actually get it." But none of that is effective with dementia.

However, once we take a look at the brain of someone with dementia – in your case, your mother's Alzheimer's disease – it's easier to see that it's a physical illness that no amount of "trying harder" can cure. Rather than dealing with dementia as a *character flaw*, we must encounter it as a *medical condition.*

You've no doubt read about Alzheimer's in print or online and seen that early on starts to damage the short-term memory and other functions handled by the hippocampus, which is a very small region of the brain. (Don't sweat it if you don't know it – you can forget it in a minute; I'll tell you when.)

But what about the rest of the brain? We don't usually talk about how dementia changes the brain in the long term.

As I said above, it's more accurate to think of dementia as *brain failure* than as just *memory loss.* You can understand why this is more accurate by looking at the drawing that compares, on the left, half of a healthy brain with that, on the right, of a brain damaged by advanced Alzheimer's. Although both brains were from people about the same age, the brain on the right is about *two-thirds smaller* than that on the left. In concrete terms, this means that while the brain of a healthy older person weighs about *three pounds*, the brain of an elderly person who has advanced Alzheimer's can weigh as little as *one pound.*

Frankly, when I consider the tremendous brain damage that Alzheimer's causes, I'm amazed that people with this condition function as well as they do. The above drawing makes it obvious why Alzheimer's isn't just about memory loss and damage to the hippocampus. *Alzheimer's undermines almost all aspects of a person's life.* (That's it – we won't talk about the hippocampus again!)

Eventually, brain failure's damage is so extensive that the brain can no longer perform even the minimal functions necessary to keep the body alive. This is why we say that *all types of dementia are fatal.* This doesn't mean, however, that your mother *will* die from dementia. She might instead eventually die *with* dementia, which means that her death was actually related to *some other injury or illness.*

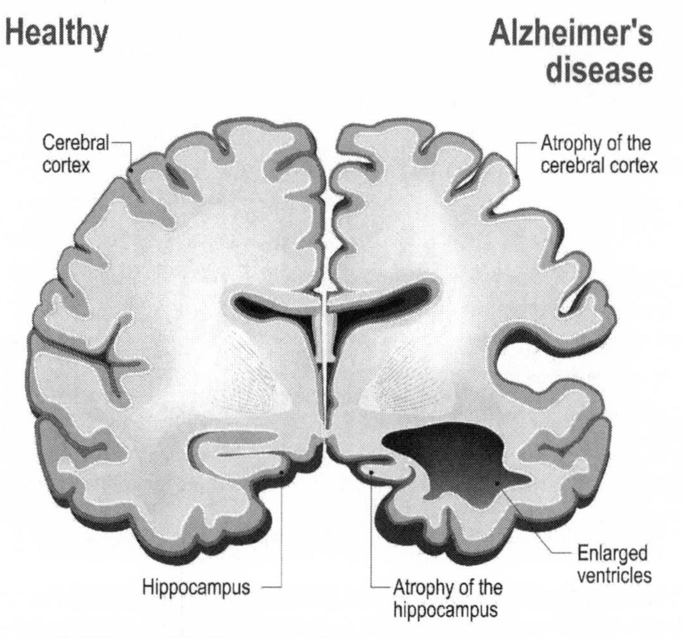


**Women and Men with Dementia**

By now you're probably wondering why I keep referring to your mother living with dementia. I'm guessing you're female, too. This is because two-thirds of family caregivers are female. Women also care more hours for those with dementia, as well as those with worse behaviors. Since we women are trained from the start of life with our baby dolls to take care of others, caring for a family member with dementia falls more easily to us. Yes, more men are stepping up caring for children, elders, and going housework, but there is a learning curve that adult women got through decades ago. I heartily applaud you if you're a man learning these skills!

Two-thirds of people living with dementia are also female. This isn't because being a woman results in a higher risk to develop dementia. It's actually because women live longer than men. Because risks for dementia increase with age, we're seeing the natural outcome of there being more women living with dementia.

**"Who, Me? Dementia? Never!"**

What frequently drives family caregivers crazy is witnessing their loved one behave and function very differently from how they did before dementia. Yet, very often, the loved one simply doesn't agree that something's wrong with them. But this isn't a display of combativeness or stubbornness: dementia causes many folks to lose the ability to see themselves objectively. They lose self-awareness and insight, which tends to lead them to believe that *you're* the one who's suddenly developed the worrisome problem. Don't try to convince your mother otherwise. It's fruitless and, in the next chapter, you'll learn why it violates Key #1.

If you'll keep in mind that your mom with dementia lacks the ability to analyze or observe herself, you won't waste time or energy trying to get her to admit she's got a problem. You'll also keep from getting stuck by the fact that she doesn't see her own behavior as alarming or lacking, although you can see these changes from a mile away. This is just how the reality of how dementia is.

**Can't Get (or Stay) Started**

People with dementia have trouble getting started – with everything. They have what I like to call *dementia inertia.* If, perchance, they do get going, they have trouble persisting in activities until they're done, including eventually ones as basic as dressing, making a sandwich, or washing laundry. They stop doing things they enjoy because they can't get started and can't figure out how to see the activity or project through to the end. They have trouble remembering what the first step is, they can't reason out what it should be, and they're no better at organizing the rest of the steps that lead to completion. Does the baloney get put out on the counter before or after the sandwich bread? Do two slices of bread go next to each other on the bottom of a sandwich? When does the mayonnaise go on? The problem here is called *sequencing*: putting steps in the appropriate order.

Often, your mom might be better at sequencing independently if you pre-stage the steps of an activity for her. This is what Monty's wife, Martha, learned below.

> Monty couldn't get dressed himself in the morning. Oh, sure, he'd put on each piece of clothing. But the underwear ended up on top of the trousers, and he'd try to put on his pants *after* he'd put on his shoes. Monty's attempts to dress himself were stymied by his *sequencing* problems. While his wife, Martha, could have helped Monty dress himself in the right order, I was certain that Martha's life would be more agreeable if I taught her a self-sequencing system that Monty could use to dress himself.
>
> Martha learned that, before she made breakfast, she could make the morning go a lot more smoothly by stacking Monty's clothes for the day into a pile on the bed: on the very bottom were his shoes; on top of the shoes, his socks; on top of the socks, his trousers; and at the pile's very top, his underwear. When Monty dressed, all he had to do was take the pile's top-most item and put it on, then the next thing, working his way down to his shoes. Now Monty's underwear is never on top of his trousers, and he feels happy he can dress himself. Martha is happy, too, because she has fewer things to do for him, and because dressing himself makes Monty so cheerful for the rest of the morning.

**They Say and Do Shocking Things**

Early on in her dementia, your mother started to lose her inhibitions, or started to become *disinhibited*. Her *disinhibition* allowed her to say and do things she'd have never done or said before her social filters and self-control disappeared. Disinhibition can be so upsetting to family caregivers that it becomes a *Very Big Problem.* Mom starts cursing like angry sailor, for example, although she used to be verbally meek. No matter how much the family scolds and pleads, as mom's disease progresses, the changes in her brain intensify her disinhibition.

But going from being verbally meek to cursing like an angry sailor isn't, I don't think, a change that poses danger to anyone. While it shocks and rattles you to hear your mom curse up a storm, it's not the sort of behavior you ought to fuss over. Public nudity in windy sub-freezing weather, on the other hand, is precisely the kind of potentially dangerous behavior that requires your immediate and thorough management.

While your mother's growing disinhibition may make you increasingly uncomfortable, know that it's not *all* bad news. After decades of socialization, many of us feel like shedding some rules as we age. We *want* to reconsider the differences between what we were told we ought to think and do when we reached this point in our lives, and what our own hearts tells to think and feel when we reach it. This is a *liberating* experience, and if you don't scold or get upset, your mother with early- or mid-dementia might just end up enjoying pushing the boundaries that perhaps limited her most of her life.

#1 of the things I most love about working with older people is that they've reset a lot of their priorities. They've decided what's most important to them, and the heck with the rest. They explore new things to do and ways to be. This is *a little* disinhibition. The disinhibition that develops in *later dementia* is not necessarily of the feisty-little-old-lady-in-a-purple-hat variety. But in early dementia, disinhibition can sometimes be a means of achieving happiness and well-being. It helps release the natural creativity that many adults have pushed away for decades.

> After long making her living as a writer, Denise developed vascular dementia. And the last years of her writing career, when a small amount of dementia had developed, were years she declared her very happiest. She dared to take on subjects of interest that she

> felt wouldn't haven't been well accepted earlier in her career.
>
> Unafraid of being laughed at by her "serious" readers, she started to write funny doggerel for the first time. She had been a very private person, but she wrote the memoir her many fans had always wanted. Though her use of words became less sophisticated and more direct, her readers and students loved her work all the more. They felt that her writing late in her career was her most accessible.
>
> Because her public hadn't picked up on her symptoms, they were shocked when, following her memoir's publication, she announced that she was retiring due to dementia. She said, "I wish only that I'd said these things sooner. I don't know why I held myself back for so long. I guess I thought, *What would Mother say?* I'm glad I took the chance. It scared me, but it's been great."

Yes, we *can* sometimes thank dementia! I recommend you do so whenever possible, because you've no doubt cursed it, too.

It's important that you know your mother isn't deliberately trying to get your goat by doing and saying things she never would have before. She no longer can adhere to the boundaries established by your and society's expectations. She's may not even able to see these limits no longer exist for her. It's not her *fault*, and it's not a *character flaw*. A bit of it has to do with aging, and a lot of it, with having a brain disease like Alzheimer's.

**Don't Try to Change All Unwanted Behavior**

Once you understand The Four Dementia Keys, it would be very tempting to try to eliminate every bit of unwanted behavior in your mother with dementia. Every irritating, annoying, bizarre, inappropriate, embarrassing, crazy thing she does, you might come to feel is in your power to exterminate.

Here's a word of advice: don't try this. Not all behavior you don't like is important to try to change. After all: you probably didn't like all her behavior *before* she had dementia. Likely you annoyed your mother, too, because family members have been annoying, irritating and embarrassing one another since the dawn of humanity. The irritation didn't start with dementia, and it's not going to end with it, either.

There are some things you just need to let go of, or you will not be able to use The Four Dementia Keys correctly. People who constantly try to change others' behavior are unhealthily controlling, and that isn't going to help the wellbeing of your household. Work to *change only what you must* to keep the person with dementia happy and safe, and you from losing your ability to cope on a daily basis.

You'll need to know which behaviors you *do* need to change, and which ones you should let be. My recommendation is to focus on behaviors that are dangerous to the person with dementia or to others. This includes physical and emotional safety, so behaviors that are significantly traumatizing to others should also be the focus of change. But be honest with yourself: just because it's annoying doesn't mean it's significantly traumatizing. Be selective and intentional about targeting behaviors for change. The "how" of making those changes is what The Four Dementia Keys are all about.

**Mom was a "Difficult Person" Before Dementia**

Some folks have a difficult time with relationships before they develop dementia. Controlling or neurotic...self-centered or an addict…and so on. Such traits can become even stronger with dementia. Whatever their most unpleasant and unhelpful coping mechanisms have been, they lean on them even more in any type of crisis, including when they have dementia.

The methods in this book can still help you care for your mother better and with more sanity even though she might have been a difficult person decades before she had dementia. However, her underlying personality cannot be erased.

Being a difficult person isn't unusual for someone with dementia. Research shows that the personality trait called "neuroticism" is significantly higher in people with dementia than folks without it. I don't mean once they develop dementia, but for decades before their first symptoms, including lifelong. In terms of personality theory, neuroticism means that many folks with dementia tend toward knee-jerk reactions, high emotional instability, a habit of negativity, poor self-control, and a low ability to manage their emotional stress. They also tend to complain frequently. These are often the folks who found it difficult to sustain healthy relationships before they had dementia. You already know how this applies to your mother – she's been in your life since before you can remember!

This is important because you need to have realistic ideas about what we can accomplish through improved dementia handling practices.

**Dementia Doesn't Get a Lot Worse Overnight**

Yes, people with dementia have ups and downs like everyone else, and some days they'll be sharper than others. But confusion, forgetfulness, bizarre behavior and other symptoms of dementia don't suddenly appear or get a lot worse overnight. Or over a few hours or days. If your mother's symptoms were to suddenly become much worse, there would be one (or both) of two reasons for it, either of which require urgent or emergent medical attention.

1. It's a stroke. This can include TIAs (transient ischemic attacks or "mini-strokes") or larger-scale strokes.

2. It's delirium. Delirium is a condition that can be very dangerous if it occurs during a hospitalization.

If your mother should suddenly develop much worse dementia symptoms while at home or out and about, it will need to be your judgement whether you call 911, go to Urgent Care, or call her physician– one size does not fit all; you're there in the moment to decide, and I'm not.

Though people without dementia can experience delirium, folks with dementia are especially prone to it and have a fifty percent chance of a delirium experience when hospitalized.

Delirium is a sign that there is some underlying medical crisis going on that needs to be found and treated. This medical issue can be an infection, injury, following surgery, a metabolic condition, drug reaction, addiction withdrawal, and more.

The most common in-home trigger for delirium is a urinary tract infection. There might not be complaints of burning with urination, though there might be more frequent urination. What you might see instead is significantly more confusion, aggression or withdrawal, hallucinations, incontinence that may be atypical for her, falls, or any other type of sudden worsening of or new dementia symptoms. Once your mother has been treated by her physician for the infection, these new symptoms (or suddenly-worse previous symptoms) should generally disappear. This is because the real problem was the infection causing the delirium.

There is not yet direct treatment for delirium, so the underlying illness must be treated and patience is required for the delirium to run its course once the medication treatment does its job.

Delirium in the hospital is a different situation; it can be both more destructive and dangerous. People without dementia who have a bout of hospital delirium have an 800 percent elevated risk of developing dementia. Afterwards, there can be either fleeting or long-lasting muscle weakness or worse cognitive impairment. Despite how vital it is to identify the presence of delirium, up to 75 percent of delirium cases in the hospital are missed. If the person has gray hair, hospital personnel will often chalk up delirium behaviors to *dementia* – whether or not the chart says dementia is present.

This is, in my opinion, one reason people who experience delirium while in the hospital have a significantly elevated mortality rate for the following year; they have had medical crises that are ignored and misconstrued as dementia.

It's important to tell hospital personnel treating your mother if she isn't acting like her usual self, and what is different. You know her best, and you're needed to prevent the situation where the symptoms of delirium are ignored, blamed on her age, or assumed to be part of her normal level of dementia…even if she didn't have dementia before she was hospitalized!

**Fast Summary – Chapter 3**

- Dementia is a disease, not a part of normal aging
- Dementia is an umbrella term for dozens of diseases that cause cognitive impairment.
- All Alzheimer's is dementia, *but not all dementia is Alzheimer's.*
- Alzheimer's comprises 60% to 80% of all dementia diagnoses.
- Dementia is *not* the same as memory loss.
- Some types of dementia don't include memory loss; some types of memory are preserved even while others are lost.
- It's more accurate and useful to refer to dementia as *brain failure.*
- Each type of dementia has its own unique set of symptoms and progresses along its own unique path.
- Each person with dementia is unique: "If you've seen one person with dementia, you've seen one person with dementia."
- Everyone involved with someone with dementia should use the same methods for care.

- People with dementia say and do things they never would have before dementia.

- Don't try to change ALL behaviors you don't like in a person with dementia.

- Folks living with dementia have trouble starting, continuing, and finishing tasks on their own.

- Dementia doesn't get worse suddenly unless it's a stroke or delirium

- This book's Toolkit contains helpful documents that can be downloaded from www.FromCrisisToCalm.com.

# 4. INTRODUCTION TO THE FOUR DEMENTIA KEYS

The amount of information on best practices for behavior handling is vast and rich. To help you learn about them more easily, I have organized these practices into four principles called The Four Dementia Keys. Because we're focusing on calming your crisis, and because solutions work their charms quickly and effectively with many (if not most) people with dementia, I've selected only the most powerful to teach you.

**What You Bring to the Table**

If you want to turn your crisis around, you'll need to bring some things to the table in addition to the approaches I teach you.

The first is a willingness to try these approaches not just once or twice, but consistently over a period of time. You need time to work with what are concepts probably very different to what you've used before in caring for your mother with dementia. You *will* make mistakes but, with practice, you *will* become more adept.

Practice is said to make perfect, but I can assure you that you will not be perfect in how you provide behavior handling to your family member. However, if you persist with this program, your situation should *improve*. It's important to let go of beating yourself up when you make momentary mistakes in care, because such mistakes are inevitable. Besides, you don't really have the time or energy for harsh self-criticism. You learn from your errors and move on: mistakes are an expected part of any learning curve.

You also need to bring with you a willingness to let go of any care approaches you're using that turn out *not* to be best practices. These are likely ingrained as habits by now, so replacing your old tricks with new ones is going to take some attention and intention. But don't worry. Despite the silly saying, we can still learn new ways of care regardless of our age, which is why this book is meant to coach you over a period of weeks.

Some of these out-of-date behavior handling approaches used to be taught as the right way to do things. Although we now know they're not the best methods, they're frequently still taught, often by the example set by long-time professionals. But despite their longevity, these old methods actually can make things *worse*, not better. These must be discarded. Through the course of this book, you'll learn the necessary specifics about what to leave behind as you replace these incorrect practices with much more effective ones.

Some family members will cling to their out-of-date practices out of pride. They feel that changing what they do is an admission that they were wrong in the past. Theirs isn't the best way to think about making changes that improve care. Instead, I remind all my students, clients – and *you* – of what I said at the end of Chapter 1:

**I have been doing my very best.**

**Now that I know better, I will do better.**

**New skills take time to develop until they become automatic.**

**I forgive myself, and let go of any guilt or anger I have.**

**I forgive myself for not being a perfect caregiver.**

To prevent family members from digging in and insisting that they're providing care just fine, be careful in how you introduce them to adopting these new care practices. I'd recommend that you read to them from this book, although perhaps not including this paragraph. We all need to maintain a sense of dignity, and making others in the family feel that they're "wrong" is not going to help them adopt new ways.

You'll notice that The Four Dementia Keys repeatedly talk about using the capabilities your mother still has to calm your family's dementia trouble. This may be a very different way of thinking about your mom. Typically, our society focuses on the disabilities of people who live with dementia, not on what they are still able to do. For example, it's common knowledge that people with Alzheimer's develop major memory problems.

It seems to be a secret that in dementia, certain types of memory remain intact a long time, even until the very end of life. The Four Dementia Keys recognize that still-intact abilities are exactly where the most powerful interventions live.

**What if Four Dementia Keys Aren't Enough?**

So, what happens if The Four Dementia Keys don't calm your crisis? What does it mean? Here are several possible reasons, and recommended next steps.

**We haven't addressed the real root of the problem.** Your mother needs a different intervention to calm this crisis than is in this book. You can use the principles behind The Four Dementia Keys to try to understand and address the underlying situation. You can also turn to the resources in Chapter 11 for further help.

**It can be hard to correctly judge ourselves.** We're so deeply embedded in the day-today of our lives that we sometimes can't see situations clearly. Even positive change can quickly become invisible to us. I recommend you think back to the day you bought this book, in the hours before you opened the front cover. Compare that with today. What has changed? What were your biggest complaints then versus now? Maybe there's been more improvement than you thought at first.

**You need help learning and integrating the Four Dementia Keys.** This is a further aspect of the trickiness of self-observation that I mentioned above. It's not easy to see for yourself how well you're using The Four Dementia Keys. Often, our most negative habits are invisible to us. Ask a fair and constructive person – *not* someone who loves to cut you down – to observe you and give feedback. They'll need to understand The Four Dementia Keys to give you a worthwhile reading. Again, you can turn to the resources discussed in Chapter 11 for help in this area if you don't have this kind of fair judge to lean on.

**Wrong timing and pace.** You need to ask yourself if this is a problem you really need to address *now*. Are you trying to address too many things at once? Are you trying to remake *everything* at the same time? Remember: only *dangerous* or *highly traumatizing* behaviors should be changed. Also, you might want to limit the areas of simultaneous change to one or two. This is exactly why I have this book broken up into six lessons, so you focus on proficiency in a limited number of areas. Slow down. This crisis wasn't built in a day. Its full dismantling might just take some time.

**The person doesn't like and/or feel safe with you – or *vice versa*.** I don't know about you, but I don't love all my family members. In fact, I dislike several of them. With families, we develop baggage that in some cases defines the entire relationship. This very much impacts the way care is both offered and received. Maybe one or both of you just can't get past the hurt or dislike you have for one another. The middle of a dementia crisis is likely not the time when either of you will be turning around your relationship. In fact, that ship might have already sailed.

The need to provide and receive care can potentially bring family members closer and heal previous rifts. But in some cases, it's just plain unwise for certain people to spend intense time together. There may be days, for example, when you and your mother are like oil and water. Your very presence is an irritant that triggers dementia behaviors. When this happens, you maybe aren't the right person to provide one-on-one care for her. There are many other ways to help that don't leave you two alone together for hours on end. See

Chapter 10 for more on other roles you or someone else in the family might be better suited to fill.

Here's a reminder that Dementia Key #1, the subject of the next chapter, is the most powerful and far-reaching of The Four Dementia Keys. Every Key that follows is served up with a big slice of Key #1 on the side. So, take some time learning how to put Dementia Key #1 into practice with your mother. If you teach nothing else to your other family members, Key #1 will take you all the farthest distance together.

## Fast Summary – Chapter 4

- You will not use The Four Dementia Keys perfectly; accept your imperfection.
- There is a learning curve with improved behavior handling skills; don't get discouraged quickly.
- The Four Dementia Keys can improve most situations, but in some cases, they may need to be supplemented with other interventions.
- This book's Toolkit contains helpful documents that can be downloaded from www.FromCrisisToCalm.com

# 5. DEMENTIA KEY #1: NURTURE EMOTIONS

Dementia Key #1 is the richest, most powerful and elegant of the set of Four Dementia Keys. It's so important to calming your crisis that you will see threads of Key #1 woven throughout the rest of The Four Dementia Keys and the remainder of this book. This is why I'm teaching it first! Over and over again, we will return to the concepts and practices I'm going to lay out here in Key #1.

**Folks with Dementia Have "All the Feels"**

People with dementia, especially with Alzheimer's disease, have fully intact emotional abilities. The emotions they experience have meaning and are reactions to either *external* conditions, or to some type of *internal* event, sensation, thought, or old memory. They often have *big* emotions that are filled with distorted ideas. The way they emotionally understand and respond to their inner and outer world can be filled with confusion, misunderstanding, and fractured logic. This is, after all, the world of feelings – no one's emotions make complete rational sense, especially to others. What's more, people with many types of dementia lose their ability to use and understand language, and as their caregivers, their over-the-top emotional displays often leave us speechless.

Because you know your mother more than well enough, you can often figure out why she feels the way she does. Not always, but you will know why consistently enough to let you respond in ways that prevent or calm crises.

People with dementia, particularly Alzheimer's, are able to experience every one of the emotions you and I experience. Those emotions are easy to spot if you're observing them. Sadness looks sad. Anger looks angry. Happiness looks happy, and so on.

**She Knows What You're Feeling, Too**

If people with dementia were, long before the onset of their disorder, able to accurately read others' body language and understand their emotions, they'll still be able to do so...like champs. Your mother closely observes the central person(s) in her life for clues about how she should feel at any given moment. She does this so she can, despite her great confusion, understand if things are going well, or if something scary like danger or upset has the upper hand. You are a barometer for her: if you look happy, things are good. If you're tense, something bad might be up. If you're downright unhappy, well, that means things are really not going right. She'll feel and act accordingly.

Your emotions, therefore, become her emotions – often magnified times 10! If you're upset, she will often not only mirror your upset, but blow it up larger than life. This can happen before a word has even been spoken: one look at you, and she knows how you feel. She'll respond quickly and with intensity.

In exactly this way, it's typical that caregivers trigger unwanted behaviors. It's not intentional, of course. I know you're not looking to stir up trouble. But it's the lack of awareness of how your body language is reflecting your inner state that unleashes a large amount of dementia behavior. Since your mother is losing her ability to control her emotions, or tailor her behavior to social expectations, you'll quickly see an outburst.

If you have a repeated problem with, say, getting your mother out of bed in the morning, your body language problem can dramatically complicate matters, just like it did in this family, below.

> Nancy was not getting out of bed in the morning without a struggle with Humphrey, her husband, who provided her care. After a

few days of this situation, Humphrey started becoming unhappy and anxious thirty minutes before he needed to get Nancy up. His stomach would start to churn and his mind would go over all the blowouts he had with her over the last few days. He'd get pretty worked up and, if you were in the room, you'd notice he was grinding his teeth in a grimace. His hands were clenched into fists. Even before his first try at getting Nancy up, he felt both angry and defeated. All he had to do was step into the room where Nancy was sleeping and, before Humphrey had uttered a word, and she'd start resisting getting out of bed – "Get out! Get out of here! Stop it! Stoooooooop!!" If Humphrey got near her, she'd try to scratch or hit him. His scowl! His tight voice! Nancy didn't know what the problem was but, with the way Humphrey looked and sounded, there was no way she was cooperating with him. Poor Humphrey had no idea why they were having this struggle. The last thing he would have guessed was that *his body language* was what started it.

When Humphrey and I worked together, he learned to let go of yesterday morning's debacle and put on a pleasant face. "Good morning, darling! I hope you slept well. You look rested. I brought you a hot cup of coffee, just the way you like it. Would you like to sip it in bed?" went over much better. He didn't even demand that Nancy get out of bed. Through his body language and speech, and by giving Nancy a cup she could hold, Humphrey gave her only love and warmth. And only after they'd had a chance to chat a bit, he'd bring her a robe and slippers, and offer to help her out of bed.

> He didn't say, like he had in the past, "I hope that today you're going to be more cooperative than the past few days. I don't have the time for that crap, and I'm not going to put up with it anymore." He knew Nancy couldn't remember a thing about the past few days, much less a specific incident he was referring to, so why bring it up? Blaming and living in the past, as every couples' counselor on the planet will tell you, is justifiably interpreted as picking a fight.
>
> In fact, Humphrey didn't even ask or tell Nancy to get out of bed. He just brought her robe and slippers and offered his hand to help her rise. She felt adored and cared for by her beloved, and she wordlessly got out of bed without a hint of a fuss.
>
> But Humphrey did even more that was correct in this interaction with Nancy. There's actually research about the effects on a person who is handed a cup of liquid from another person. The evidence is clear that, if the liquid in the cup is hot, the recipient feels greater *trust* for the giver, even between strangers. If the liquid is cold, *distrust* is increased, even between strangers. That hot cup of coffee Humphrey handed Nancy wasn't just a fragrant, delicious morning pick-me-up. It was also a trust-building insurance policy.

Happily, the flip side of the way that unguarded negative emotions trigger dementia behaviors and upset is that, if you choose to wear positive emotions, you get a positive result. You'll see your mother reflect your emotions right back at you, often larger than life. And if you pay attention to your body language and choose to express feelings that *you* like to feel, your mother will respond with happiness, delight, warmth, gratitude, or whatever emotion you showed her when you walked into the room.

**Learning the Dementia Two-Step**

Two changes that Humphrey made in turning around his morning routine with Nancy are the most important ones. He did the **Dementia Two-Step**, which means he learned to dance with both his and Nancy's feelings in a simple but really powerful way.

> **Step One:** He examined and **Left His Baggage at the Door**, adjusting his body language to be positive and loving.
>
> **Step Two:** He stopped and spent a moment to **Share Some Sweetness** with her, without hurry or another agenda… just *sweet* sharing.

Immediately, this changed the mood of the entire morning routine. Nancy reflected right back to Humphrey the feelings that he was showing to her.

From now on, try to be like Humphrey. Start *every* care task with the **Dementia Two-Step: Leave Your Baggage at the Door** and **Share Some Sweetness**. *Do not proceed with any care task or interaction until you have completed the Two-Step together*, and you're both feeling more relaxed and connected to one another.

Also, do the Dementia Two-Step *before* making any requests of a person with dementia, *before* interacting with them in any way, and whenever things are suddenly going wrong – even if you don't know *why* things went pear shape. Just stop, step back, and Two-Step forward, *together*.

To Leave Your Baggage at the Door, take a self-inventory of your body language. How is your posture? Your tone? How quickly or slowly are you speaking? What are you doing with your hands and face? Do they express, "I'm a relaxed and happy person to be with," or "You pain in the butt – I'm not going to put up with your pushback any longer"? Make sure you're showing your upbeat side, and your mother will be more upbeat, too.

There are many ways to Share Some Sweetness. In fact, there are as many ways as you can think of, and more. Whatever the task you want to accomplish, it only needs to take a minute or two to share sweetness first. As long as it's pleasant for at least the person with dementia, safe to do, and a shared experience, anything goes. Some choices might be:

- Listen to her favorite music together

- Sing together or take turns

- Notice the birds together out the window

- Give her a hand massage

- Swap silly stories or jokes – they don't need to make sense

- Drink a beverage together

- Play with a pet together

- Hold hands

- Look at and talk about her favorite book, picture, or video clip

- Ask her to tell a favorite story...yes, yet again...and show pleasure at her enjoyment of the telling

- Walk, dance or stretch together

- Together listen quietly to the sounds around you

You get the picture…

I hear you saying: "Thanks, just what I need...more things to do! I don't have time for this Two-Step business!" I understand what you're saying. You're already feeling so overwhelmed. However, each of The Four Dementia Keys will overall *give you more time back than they take from your day*. This is because they can help prevent the dementia behaviors that take so much time to calm before you can get back to doing useful care tasks. You might be spending so much of your time now dealing with emotional meltdowns and care refusal that it's hard to get anything done. Approaches like the Dementia Two-Step can actually make you more efficient and less fatigued because care will go more smoothly.

People with dementia know when they're being treated like an unpleasant chore – we all do, really. *No one wants to be treated in an impersonal or dismissive way.* You perhaps didn't know this matters – a lot – to people with dementia. Now that you know, I'm certain you'll do things differently. You'll nurture positive emotional experiences for your loved one living with dementia by doing the Dementia Two-Step many times through the day. You'll very likely feel better as a result, too.

## Highly Sensitive to Environmental Emotions

The emotions around them impact people with dementia more powerfully than many other folks. This is why *everyone* in the family should be using the Four Dementia Keys. Your mother will "catch" any feelings around her and reflect them back with her own feelings and behavior. Again, her reflection of your emotions can be several far more intense than the emotions you expressed when you walked into the room.

For example, if there are arguments going on among your other family members living in the house, and they're getting on each other's nerves, the person with dementia might become irritable or upset, too…*only more so.* And a person with dementia's irritability can increase others' irritability, which in turn escalates everyone's anger. Before you know it, there's yelling and screaming. It's a downward spiral that can get out of hand easily. And this is how things can quickly escalate into violence from someone with dementia.

If others in the household are cheerful, your mother will "catch" that vibe, too, and reflect it back, usually magnified. In this case, everyone else is going to feel happier as a result, which makes the person with dementia even happier, and so on. An upward spiral occurs, with good moods leading to even better ones.

Your mother with dementia is losing her social filters as well as control of her emotions. This powerful combination means the emotions in the room can easily trigger behaviors that get *everyone* upset. If you don't like how your loved one is acting, then look at the way others (including, but not only, you) around her are emoting. If you want to see her calm and happy, others must be that way when they're with her. If everyone doesn't intentionally set the emotional tone of their interactions, then their fatigue, worry, hurry and exhaustion might just be setting the agenda – with unpleasant results.

## Their Feelings are Consistent and Long Lasting

I need to make another really important point about the emotional capabilities of people with dementia, especially Alzheimer's. The feelings someone like your mother has about you, other people, places, and things, remain quite consistent over time. Though she likely doesn't remember why, if she never liked her neighbor to the left, when she sees him now, the same feelings arise. This is true even if she no longer recognizes him or doesn't recall that they're neighbors. Her feelings are *consistent.*

If she loves her neighbors to the right but can no longer recognize them and doesn't know who they are, she'll *still* have positive feelings when she sees them. She will have the same reaction over time despite her failing memory. (Remember how I experienced the same when I had an injured brain? I had no idea who people were when I bumped into them, but I could trust my feelings would accurately reflect history I couldn't recall.)

As a result of still-intact emotions, many people with dementia still suffer the long-term effects of past trauma they can't recall or talk about. This could be the case with your mother. This might be true *whether or not* you know she's experienced significant trauma: there might be secrets that outlasted her ability to recall or relate them. It seems counter-intuitive that a person with even severe memory problems can carry with them still-open scars they have no memory of, unless you know their feelings remain consistent regardless of memory loss.

Another way emotions are consistent for your family member with dementia is that any feeling state will last until something else happens that creates a stronger and different feeling state. If after lunch your mother visited with that favorite neighbor and she had a great time, her mood will be a good even after the neighbor has left and she's forgotten the visit had ever happened.

The same is true of that fight around bathing you two have had, one that maybe escalated one day to pushing and hitting. Though Mom quickly forgot the incident itself, she'll likely remain easily upset, nervous and agitated for hours, if not for a day afterwards. The next time you even mention shower time, these feelings might come flooding back and her behavior can quickly pick up where things left off.

## Your Loving Relationship is A Powerful Care Tool

Whether a happy, sweet relationship or a sad, contentious one, your relationship with your mother living with dementia is what sets much of the foundation of how daily life goes on between you. The strength or weakness of the relationships in any household colors life within its walls, dementia or not. But for people living with dementia, this is magnified due to their emotional sensitivity and tendency to have bigger-than-life emotional reactions.

Lifelong relationships often contain emotional baggage. Some families do beautifully facing their conflicts with honesty and love, with a goal of healing for all. Others never face – much less settle – their problems. The latter can create mountains of ongoing hostility for years, even generations.

The late United States Poet Laureate, Maya Angelou, wasn't an expert in behavior handling, but she was a genius of the human spirit. She said, "I've learned that people will forget what you said, people will forget what you did, but people will never forget how you made them feel."

The Angelou quote above is true of your family member with dementia, who might forget your name and who you are to them. ***But they never forget how you make them feel.*** Please go back and read that sentence again. This idea is *central* to calming any dementia crisis, and we will come back to it again and again. While your mother may no longer recall she's ever met you, she'll have instant access to her feelings about your relationship. If you've been close, she'll be happy just seeing you. If you've always had a contentious relationship, just seeing your face might make her uncooperative, angry, sad, or afraid.

**"Calling Up the Love:" Transformed by Sweet Feelings**

It can be helpful to practice "**Calling Up the Love**," which means, before being with your mother, you connect to the love and warmth you have felt for her. Use this consistently, especially when you feel you can no longer recognize her because she's so unlike how you've always known her.

Calling Up the Love is a tool you can consciously use to keep your interactions positive. Sometimes, family caregivers get so exhausted and frustrated that they stop offering care from out of their love for the person with dementia. Instead, resentment, anger, obligation, and fatigue take over. As part of practicing the Dementia Two-Step, center yourself in the love, affection, caring, and warmth you have for this person before you start any care task or interaction. Your family member will feel it, and you'll do everything differently in both subtle and big ways that will make things go more smoothly.

This is an important technique, as your love will be transmitted to the person with dementia even without anyone speaking a word. We're transformed when we're centered in love, and your mother on some level will notice, and she'll feel more love and warmth for you, too. Give it a try – it's not easy to do this when there is conflict, but the results can be truly worthwhile. Be patient and work with this method over time for it to come more easily to you.

**Be Here, Now**

Your mother can be sad and anxious when she worries about the past or future. When she's able to be present in the moment, she can be more peaceful and happier. The funny thing is that it's *exactly* the same for you and me. Learning to be present in the here and now is beneficial for *everyone* to cultivate on a daily basis. This is at the center of the Mindfulness movement that's permeating more aspects of our culture every day.

Interestingly, there's now a large body of research about the emotional and physical benefits of a variety of meditation practices, including Mindfulness-Based Stress Reduction (MBSR) Meditation and Kirtan Kriya Meditation where there is dementia present. Research shows that people with dementia can be taught these methods successfully, and that positive results for both family caregivers and their loved ones with dementia can be achieved. These benefits include a variety of quality of life measures, less anxiety and pain or the perception of pain, fewer stress symptoms, better regulation of emotions, higher self-esteem, increased feelings of gratitude and appreciation, better relationships, improved immune response, and more. And there are no known negative side effects!

There are many tools to help you and your mother practice meditation. Following guided meditations can be helpful for people with dementia who, if they meditate in silence, will often forget what they're doing and wander off. I've included a document in the online *From Crisis To Calm* Tool Kit that's part of this book that lists several links to both MBSR and Kirtan Kriya, since these forms have been researched with people with dementia. Download this document at www.FromCrisisToCalm.com. And then meditate together regularly – you could both feel better for it.

Slowing down and being more present is important because people with dementia simply *cannot* be rushed. Being hurried likely results in increased confusion and emotional upset. Since your mother with dementia picks up on your emotional state and reflects it back to you, often magnified, if you feel and act rushed around her, she can end up having an emotional meltdown.

Luckily, one of the best gifts family caregivers can receive from their loved one with dementia is the opportunity to be together at the family member's slower speed. So many of us need to s-l-o-w d-o-w-n and stop feeling pressured to rush and fit more and more into our day. For your *own* wellbeing, intentionally and regularly spend some unhurried time at dementia's slower pace.

**"She's Not Like My Mom Anymore"**

Changes in personality are a hallmark of dementia. "I don't know who she is now" is what many families say to sum up their loved one's recent most head-spinning, unpredictable behavior. Often in addition to personality changes, we see some aspects of the person's pre-dementia personality actually become more acute. It's often – but not always – the traits most difficult for you and others that seem to become more vibrantly unpleasant.

You might experience a painful sense of loss and grief when you can no longer recognize the inner essence of your mother. The really good news is that if a person with dementia is cared for using state-of-the-art-the-art methods such as the Four Dementia Keys, they can more often be seen being "themselves". Some large portion of the aggression, agitation, anger, and upset we see in people with dementia is an expression of the daily stress they're experiencing. Some people with dementia are even in a very high level of emotional crisis much of the time.

Have you ever experienced a period of great life stress? #1 that just went on and on, where it feels like your feet were repeatedly kicked out from under you and you're constantly thrown into emotional freefall? If not, perhaps you've observed this state in someone close to you. I'm talking about something on the order of a difficult, long divorce or child custody dispute. Think of the folks living at the focal point of such stress. I bet you've seen them have some personality changes, too. Did they say and do things atypical for them – maybe even erratic and bizarre? "It's the stress getting to her," we'd say, quite correctly.

People with dementia who constantly "act up" are often expressing that they, too, are feeling ongoing stress. It may be due to a wide variety of reasons, some of which we'll go into in Chapter 9, so let's leave them aside for now. Part of why people with dementia stop seeming like "themselves" and don't act "normally" is that they're highly distressed much of the time.

Frequent or high levels of distress can also be blamed in part on changes in the brain due to their dementia's disease process. Folks with dementia become disinhibited and have trouble controlling their emotions as a symptom of their brain failure. Some mood disorders, particularly depression that first appears in late life, are thought to be one of the earliest signs of Alzheimer's disease, even beginning years before any memory symptoms. There is a biological aspect to the highly negative emotions we see just under the surface with many people living with dementia.

It's possible that, as you learn the care methods in this book, you'll notice your mother acting more like "herself" as she calms down and experiences less ongoing stress. Your ability to improve your behavior handling-caregiver-game can have powerful effects!

**Solving Behavioral Puzzles**

Why does your mother do such strange things since she's had dementia? This question is central to the perspective shift I'm teaching throughout this book. You need to *always* remember that there is *meaning* to even your mother's strangest behaviors and emotions. Recall that language use becomes more and more difficult as a dementia like Alzheimer's progresses. Eventually, all your mother has left to communicate with is her behavior. By learning to decode that behavior, you'll be a step closer to understanding what's triggering it.

When it comes to unwanted dementia behaviors, you should always start with the assumption that they're your mother's way of telling you that something is bothering her, that she has a need that's not being met. Once her behavior is decoded into communication, you can then create a response that truly fits the needs she's expressed. Without that decoding, you'll stumble around in the dark, hoping you get lucky and can somehow quell the storm.

Do you enjoy a mystery story? I do, and of course I try to figure it all out as each step of the story unfolds. I certainly want to know what happened before the big reveal at the end. This is a skill similar to figuring out what your mother's dementia behavior means. There are three things you need to think about as the first steps in decoding. (We'll talk more about this topic in Chapter 9, by the way).

**1. Is she expressing a biological need such as hunger or thirst, fatigue, the need for a bathroom, or is she too hot or cold?**

These can cause all types of dementia behavior. Since these are frequent and universal needs, start decoding here. Be proactive. Because it's likely she'll not tell you what's up in easy-to-understand speech. Try toileting, feeding, hydrating, encouraging a rest, and responding quickly to changes in temperature, to see if addressing any of these factors can calm the waters.

**2. Is she in pain?**

People with dementia don't tend to complain of pain, but research shows they experience just as much – if not more – pain as folks without dementia. People with dementia don't behave the way we'd expect when experiencing pain. Your mother might not be showing some of the most obvious signs of pain that she used to, despite pain's current presence. Missing signs (including a lack of verbal reports) make it difficult to guess there's pain at the bottom of many dementia behaviors that can't be calmed as usual.

The problem your mother has with communicating pain to you or someone else is partly due to her dementia-related struggle with using language. And because people with dementia can lose an understanding of what their body sensations mean, your mother will also have trouble realizing this difficult thing she feels is called "pain." What's more, due to dementia-related neurological changes, she might not be able to locate pain in her body. She might even consistently say she's not in pain when asked. However, research shows that unreported, unmanaged or undermanaged pain is still very common among people with dementia.

Researchers gave people living in a memory facility a little stronger pain medication than they were already using to see if it would impact their behaviors. If they had no pain medication, they received acetaminophen. If they already were using acetaminophen regularly, they got something the next stronger level up, and so on. No one was drugged far beyond their current level of medication. As a result, dementia behaviors were considerably reduced, indicating that pain is often at the bottom of these unwanted, dangerous behaviors.

Though other types of drugs besides those for pain management are used to stop dementia behaviors, none of them actually do a thing to address their pain. These include antipsychotics and benzodiazepines, two classes of drugs used widely – and too often inappropriately – for people with dementia. Make sure your family member is getting adequate pain management and isn't just being sedated. Ask your family member's physician for help with this.

> Lin wasn't acting typically all day. She had some type of dementia past mid stage. Since she didn't speak often, she couldn't tell anyone what was bothering her. She was highly agitated, crying a lot and not interested in meals. She had taken a fall that morning, but she hadn't seemed injured. She wasn't limping or holding any specific spot as we humans do naturally at the site of an injury. Her family figured she had injured herself, but where? She shook her head when asked about pain.
>
> At the emergency room, Lin was very difficult to handle. The doctor gave her a sedative in order to examine her. Through x-rays, they found she had broken her right radius, a bone in her lower arm. She hadn't reported pain, nor had she held or guarded her arm as if it were hurting her. In fact, she was using that arm a lot as she flung out her hands out distress. But after a few days with her arm in a cast, taking some pain medication, and getting some rest, Lin became more her normally calm self as her discomfort faded.

### 3. Is she bored?

As they say, idleness is the devil's workshop. None of us like to be bored, and many of us have been known to act out and stir up trouble if not sufficiently engaged (including and especially me). This is particularly true when there's too little to do that feels satisfying and purposeful. It's exactly the same with people with dementia. They act out and stir up trouble, too, when there aren't enough satisfying, purposeful things to do.

Recall your mother has sequencing problems and doesn't know how to start or keep an activity going until it's done. Her relationships can feel difficult and complicated, not rewarding and sweet. She has dementia inertia, and can't start new things, or keep them going until the end. She has a short attention span and is easily distracted. Her capabilities have deteriorated, and she hasn't a clue how to adapt her former interests and hobbies to her new ability level. Day in and day out, she could be bored, and unable to relieve that boredom without help.

Her remaining abilities will also be lost more quickly if they're not used due to inactivity. If someone doesn't help her find things she can do that are safe and happy for everyone, she'll find something to do that's possibly dangerous or that you'll otherwise object to. It's up to those in her environment to help her keep happily engaged and busy.

### Living a Life of Purpose and Meaning

All human beings have a drive to live with meaning and purpose. If you or I lost our purpose and meaning, it could result in a distressing crisis until we discovered a new sense of meaning and purpose. This is very common when someone becomes seriously disabled: what is their reason for living? What purpose inspires them get up in the morning? Imagine your mother's feelings about having lost her sense of life meaning and purpose. And now think about how much trouble she has figuring out how to do even the simplest things. She can't now on her own search for, and refashion, a life meaning and purpose now. She needs others to help them.

So, ask your mother to help with day-to-day activities that she can still do, even if she's just a little bit of help, or even *no* help at all. And then thank her. "I have so much to do, I don't know how I would have managed without your help. Thank you. Can you help me again tomorrow?" can give her some sense of purpose. Ask her to help you with things you know she used to enjoy doing, even if you're really doing them only for her pleasure. Declarations such as "I love seeing you when I come home from work," can help her know she's still valued by you.

She'll probably forget what you said, but she'll *never* forget how it made her feel.

Between consistently using the Dementia Two-Step and the above three questions to begin decoding her behavior, you're well on your way to calming your dementia behavior troubles.

> Theodore, a man with Vascular dementia, would suddenly stand up out of his chair at home, grab his hat and coat, and announce he had to go. He was frantic and his wife, Francine, would go out with him for a short walk, since he was so intent on leaving. When they got back in, Francine would discover he had wet himself. She thought it was his being frantic that made him urinate.
>
> She was someone I was coaching, and in response to my questions, I noticed that these incidents happened in the late afternoon. I suggested she try Theodore on a toileting schedule during those hours: every two hours to the bathroom to urinate starting with right after lunch.
>
> Once on this schedule, he immediately stopped jumping up and insisting he had to go – the behavior was over, and so was his loss of bladder control. This whole hat-and-coat routine was because, although Theodore knew

> he had to urinate, he couldn't figure out a better way to get that need met. He might not have even realized that his physical discomfort was due to a full bladder, and he mightn't have found the bathroom by himself, either. Decoding his behavior as a sign of discomfort made the right solution quickly apparent.

**Making Choices and Having Preferences**

As adults, we get to make lots and lots of choices: from how we like our coffee to whom to marry, and everything in between. Our choices about our personal lives are often based on our preferences, preferences that are expressions of who we are, what we like, and the power we have as adults to make such decisions. The older we are, the more decades we've spent making these personal choices. In our Western industrial culture, even children are given choices and are encouraged to develop their personal preferences.

People with dementia have spent decades as adults making choices. Your mother still wants to make her own decisions. She wants her personal preferences respected. And when her personal preferences are respected, she feels that *she* is respected. It's to your best advantage, too, because respecting her preferences is an excellent way for a person with dementia to feel happier, more cared about and willing to accept care. (see Chapters 6, 7 and 8 for more about preferences and choices).

The underlying issue here is one of control. As human beings, we all need to control things in our lives. While *everything* can't be controlled, if we don't have *enough* control in our daily lives, it can be highly stressful and unhealthy, both physically and emotionally. Trying to have too much control can be deeply damaging to our relationships and health, too. There's a healthy middle ground where, as the Serenity Prayer suggests, we have the serenity to accept the things we cannot change, the courage to change the things we can, and the wisdom to know the difference.

People with dementia struggle to maintain some control of their lives, even as their brains deteriorate. Because of their compromised safety-awareness and their failing ability to reason, they can no longer safely make many kinds of decisions for themselves. Note I said *many*, not *all.* There are some decisions your mother still can and should be able to make. As a family caregiver, you need enough "wisdom to know the difference" on her behalf, to give her back the choices she can still make.

> Brad was no longer safe living at home due to his advancing dementia. His adult children hired home-care agency caregivers to be with him during the day. They tried several different agencies with the same result: Brad got ugly and drove the caregivers out of the house in the first 20 minutes, then refused to let them back in. The family is in crisis because the adult children all work and are raising their own kids. They can't spend their days caring for Brad.
>
> The family then hired an agency with a dementia care program I created. When the caregiver, Natali, first arrived, she found Brad binge watching a TV series including his favorite actor in the family room. She stood in the doorway of the room watching with him. After the episode was over, she started expressing admiration for "his" actor. "Wanna watch the next one with me?" he asked, motioning for her to sit down in a comfortable chair next to his.
>
> With that, Natali became Brad's caregiver. He always looked forward to her visits because he felt his preferences and choices were respected. He felt understood when she was there.

The decisions your mother with dementia should be allowed to make are ones that involve her personal preferences, as long as you help her to make *safe* decisions. What time she likes to wake up...how she wants her coffee...what clothes to wear...music to hear...shower or bath...what time of day to bathe...and so on. *If you take even these choices away from her, the only way she can exert control in her life is to resist her caregivers every step of the way.* If she refuses care constantly, consider this a possible underlying issue which you can address.

### "If I Hear That Story *One More Time…*"

Have you ever watched a movie you've seen and loved before? Many people have a few favorites they watch over and over again with great enjoyment. The same is true of your mother with dementia: she likes telling a few stories again and again. And again!

It's fascinating how people with significant short-term memory loss can tell stories from "the old days," often with great detail and clarity. Usually they have a few stories, typically of events approximately from ages 5 to 25, that they tell repeatedly. These stories are rarely brief, and they obviously revel in their telling.

When your mom repeats stories, it makes your family *absolutely crazy*. You've all already heard these tales 25, 50, 100 or more times. Every time she launches into a retelling of her favorite story, the response from immediate family is one of impatience, tuning-out, and interruption. *Anything* to stop the story in its tracks, which it *never* does. The story gets told anyway, and everyone ends up feeling disrespected and at odds.

I totally understand these objections to the same story again and again…and *again*. However, your mother living with dementia might not remember that she's ever told this story before, even if she's retold daily. She tells this tale often because it's an important and clear memory she can still share, while lots of her other memories are so foggy or even absent. Also, the story is repeated again and again because she likes how she feels when she tells it to a willing audience, and she wants to recreate those good feelings *frequently*.

Yet, many of us would prefer to gnaw our own foot off to escape hearing that story again. *And again.*

I know you've been doing your *very* best, and hearing those stories repeatedly can feel insulting. I've experienced this with my own family, too. Being trapped, having to hear the same stories again and again, makes most listeners feel invisible, disrespected, and that their precious time is being frittered away. Often, repetitive stories elicit a sarcastic response. Because people with dementia tend to take things literally, at face value, sarcasm isn't always easy for people with dementia to understand. Sarcasm can make a tense situation all the more so.

I suspect you see where I'm going with this by now. When you reject a story that's really important to your mother, you're setting the stage for trouble between the two of you. And that trouble is bound to spill over into the care you give her, as well as into her general mood. Is it worth it to shut your ears and heart to the story, given the possible emotional and behavioral benefits of listening?

No, I'm not suggesting you just sit there and feel tortured. I'm suggesting being strategic about things. Here are some ideas:

• Prompt your mom to tell a favorite story to someone new, such as a hired caregiver or a visitor.

• If the story begins when several other people are present, agree ahead of time that which one of you will be the one to pay close attention and be responsive to her. Take turns being the listener, freeing up others from the tedium of repetition.

• Actively solicit a favorite story when you're in your most attentive and loving mood, not when she decides to tell it. Her stories *will* be told regularly, so spark them at the time you're most able to patiently receive.

• Switch your perspective from listening to watching. Focus on your mother's pleasure at sharing the tale. Respond with the same intensity and similar feelings.

• Pretend this is a new story to you. Allow yourself to be in suspense, ask questions that prove you're listening, and show emotions as appropriate to the tale.

Whatever you do, you know the retelling of the story isn't going to stop as long as your family member can still speak. You have two choices of how to respond: in a way that improves her mood (and yours), or way that makes both of you unhappy. If she's unhappy, it's going to make things unpleasant for you, too. By now, you understand that how you engage with her is going to have a big impact on her moods, for better or for worse. It's a better strategy to help her feel good about herself and your relationship, to find a way to give her a receptive audience for her most treasured tales.

**Last but Not Least: Nurturing *Your* Emotions**

I hope using this information and practicing the skills in this chapter finds you feeling more relaxed and hopeful. While caring for your mother with dementia requires sacrifice for everyone in the household, you don't have to destroy your own health and wellbeing in the process. In fact, it's likely you'll feel better emotionally as your dementia crisis is calmed by using these improved care techniques.

However, everyone is different. Everyone's circumstances and reactions are different. Even though you get rest, if you find that you still feel crushing stress, upset, numbness, or exhaustion, you may be experiencing Caregiver Burnout (CGB), which is not uncommon with dementia caregiving. CGB isn't just being tired. It's a destructive and harmful state, both to you as well as to your mother. We'll talk about this more in Chapter 10.

**If you feel so bad that all you can think of is to run away, that your life is hopeless, or that you want to hurt yourself or the person living with dementia, please pick up the phone and call for local help immediately. Call 911 (or your emergency number) or go to a local Emergency Room.**

If you feel it's not *quite* that urgent, then call your primary care physician, a trusted counselor or member of the clergy, or some other vital source of support. Such feelings and thoughts mean you're in a crisis that needs *immediate* attention.

I know there are times that we dislike even our nearest and dearest. Maybe it's a passing feeling, or perhaps it's long term. But if your mother is someone whom you really don't like – perhaps she caused significant trauma for you, or maybe she doesn't like you for whatever reason – these feelings of dislike color every single moment you interact together. Caring for her might be making one or both of you miserable. If it were simply that you have little in common, you could likely "fake it 'til you make it" by putting on a cheerful face and eventually getting into a nice groove as the care goes well. But if you're caring for someone about whom you feel great pain (or vice versa), the care may be going poorly, which would be understandable.

Even if your mother doesn't remember who you are, she still has fully intact emotions that surface when you're there. You certainly remember her well, and whatever it was that went on between you. You may not be able to improve your relationship at this point, try as you might, because she can't participate sufficiently in the healing between you. Any time we have a lot of powerful feelings about a person we care for, that person will likely pick up on how we feel. That means that you might not be the right person to be a primary caregiver or to provide personal care.

It's also not necessary to become your mother's slave, asking only "how high" when she says "jump." These methods aren't meant to be about appeasing her so she'll just shut up and be quiet for a while. The intention here is to make her – and others around her – genuinely happier and more rested, better connected and emotionally warmed. If this isn't happening in your corner of the world, then see some additional resources in Chapter 11.

## Fast Summary – Chapter 5

• Dementia Key #1 is to nurture emotions, the most powerful of the Four Dementia Keys

• People with dementia have intact emotional lives, especially in Alzheimer's.

• People with dementia know what others are feeling through their body language.

• Unguarded caregiver body language is one of the major triggers for dementia behaviors.

• The Dementia Two-Step is vital to do before every care task or interaction, or any time things suddenly go wrong.

• The Dementia Two-Step only requires a few minutes for potentially enormous gain.

• Caregiver emotions are "catching" – show only the ones you want to see come back at you, magnified.

• For someone with dementia, their feelings can be strong and long-lasting, even if they don't remember what triggered them.

• "They may forget who you are, but they will never forget how you made them feel."

• A positive relationship is the most important dementia caregiving tool.

- Two forms of meditation have been shown to improve quality of life and behaviors for people with dementia and their caregivers.

- Changes in personality are a hallmark of dementia.

- There is meaning in all behaviors; learn the meaning to construct an effective intervention

# 6. DEMENTIA KEY #2: CUSTOMIZE CARE

The world of behavior handling isn't one-size-fits-all. Because we're each so different, no two people experience dementia in the same way. How people even with the same type of dementia decline is quite individual in terms of speed and symptoms. When it comes to our many likes and dislikes, we are again each unique. Therefore, it's absolutely vital that you customize the care methods you're learning to fit your mother's individual preferences, life history, abilities, disabilities, and interests.

You'll notice that customization is built into the most important (and most frequently used) care method, the Dementia Two-Step. You customize the first step (Leave Your Baggage At the Door) as you learn about your worst habits of body language, or the ones you'll need to consistently address in the first part of the Two-Step. For me, my shoulders rise and when I try to smile, it comes out looking more like a little grimace – these are sure signs that I'm tense and anxious. This is an area where some feedback from others is helpful, because unless you spend your day in front of a mirror watching yourself constantly (and who's got time for *that*?), it's difficult to truly see your own body language.

Share Some Sweetness, the other step in the Two-Step, is to be customized not for yours, but for your mother's interests. How you Share Some Sweetness greatly varies depending upon both of you in combination. She's your mother, so you likely know a lot about what pleases her – though there's always a few surprises. You probably already have a long head start knowing how to do things the way she likes…the ways that hit her sweet spot *just right.*

**But What Can She Still Do?**

This is one of the most important questions you need to ask as you tweak care tasks to fit your mom's needs and preferences. You'll need to identify her still-intact abilities for a few reasons. First, we want her to maintain those abilities for as long as possible. "If you don't use, you lose it" is a truism in behavior handling. Therefore, you should be doing for her *only* what she cannot do for herself. Otherwise, her abilities will decline more quickly. When you let her do what she still can do, even if she requires *some* help from you, she gets to exercise and reinforce her still-present abilities.

For example, you could brush your mother's teeth for her when she can't brush independently any more. But you could instead brush your own teeth while standing beside her, both of you with loaded toothbrushes in hand. Your mother might be able to mimic your movements just fine, doing for herself what she couldn't do without your acting as a model for her. If she's able to mimic you, it's possible she'll keep that ability for some time before she needs additional help brushing her teeth.

Not doing for her what she can do for herself preserves her dignity and pride as an adult. Even if she needs some help from you, it's not the same as needing you to do tasks start-to-finish for her. Accentuating teamwork and give-and-take are normal in a family; none of us gets through life fully on our own. Working together as a family has likely been important to how your mom's household always ran. This family-helping-family should feel familiar to her, especially if you ask her to brush her teeth with you to make sure you do a good job on your own pearly whites. Helping her do things that feel "normal" is important to her, because much of her life feels so very abnormal.

> Jack was no longer able to carve Halloween pumpkins due to his dementia, although it

> had been something he'd done with pride for decades, both as a child and with his own children, and then with his grandchildren. But he wasn't now coordinated enough to use the carving knife safely. Getting together pre-Halloween wasn't something that he and his extended family were ready to give up yet, though. They decided to change the event to one of decorating pumpkins with wide felt-tip pens. Jack's creativity was still intact and he could handle a fat marker pen easily. It was a little work to change the event from carving to drawing but, in the end, Jack and his family all had a great time together, and all the pumpkins got decorated.

If you help your mother only as much as she needs, then she will be happier, more cooperative, and capable of some degree of self-care and independence, however small that might be.

**The Little Things Matter**

It's good to go over the specifics within your care tasks to get the details to fall within your mom's preferences. Because too much change at one time can be upsetting for someone with dementia, you shouldn't change *everything* all at once to suit your mother. Make what you guess might be the most impactful changes and see how they go. Then, over time, add others. This way, changing over time can be less stressful for *you,* too.

Here are some questions to give you an example of how to customize dressing in the morning. Play off these ideas to find your mother's sweet spots. If dressing is a problem, these could enhance her good moods and sense of calm, making it easier to care for her. The answers to the below questions will probably change as her dementia progresses, too.

- Does she like to get dressed before or after breakfast?

- Does she like to dress more warm or cool?

• Does she bathe first before getting dressed?

• What colors does she like to wear?

• Are there any pain or mobility issues that impact ease and comfort while dressing?

• How easy on/easy off does she need her clothing to be?

• Does she have any skin issues that impact the clothing she wears?

• Which are her very favorite clothes to wear often?

• What clothing does she find most comfortable?

• Where does she like to get dressed?

• Does she resist undressing in winter because she gets too cold?

• Are there special stories connected with the day's clothing to chat about?

This list of questions combines all the typical preferences and habits I normally see around dressing, as well as questions about the person's abilities. You should add any more unusual questions as they apply to your mom. Once you know these answers, then you can apply them to the day's care tasks.

But don't become frustrated or criticize her if, right now, she's not really into what she once loved. Most people's preferences change over time, including your mother's. You'll need to do some experimenting to see what her new preference has become.

This approach can make everything you do with your mother easier. Her improved mood and more cooperative attitude are worth the effort, because there will be fewer emotional meltdowns and more acceptance of care.

## Stop Correcting Her

People with dementia have many unusual ideas about their history and the world around them. Many travel in time, so that on Thursday, it's 1942, on Friday, it's 1967, and on Saturday, it's 1953. They get confused about the meaning of what's in their environment, often wildly misinterpreting what is going on around them.

In response, you and your family probably argue and debate with your mother, explaining exactly how she's wrong in detail, trying to convince her by using facts and logic, criticizing and generally trying to set the record straight.

As is common, you and your family correct her over and over again, because she keeps getting the same things wrong, over and over again. It's clear that your family's efforts are completely and totally ineffective. Living with dementia, you mother cannot be argued or explained into being better oriented in time and place. (Although, in early dementia, this might have been sometimes possible.) Nor can she be successfully taught how and why she's wrong. Because all she has is fractured logic, she can't be convinced with facts and rational thinking. Correcting her is useless for fixing what's not understood.

Most of all, by correcting her, your family completely violates the second half of the Dementia Two-Step: Spread Some Sweetness. There's nothing so destructive to a relationship than constant arguing, correction, and criticism. Given how often she gets the facts wrong, your family needs help in no longer correcting your mother from sunup to sundown. Correcting is not just *useless*, it's *harmful* to all of your psyches. Starting today, you need to work at no longer:

- Correcting your mother with dementia

- Using facts and logic to convince her she's wrong

- Explaining things in different ways, hoping some approach will finally get through

- Arguing to get the facts right

Be aware that this change takes some time to achieve consistently. So many families argue all the time. They correct, criticize, and try to convince one another just about 24/7. It's what we do in our culture with family, colleagues, employees, neighbors, and friends.

I suspect you're like me – I can stop arguing much more consistently with strangers with dementia than I can with my own family. So be prepared that you will make mistakes. Likely, you will hear yourself arguing regularly. You won't believe how often you catch yourself, and you might end up feeling like a complete jerk because you're *still* correcting even when you're trying hard not to.

There's a real learning curve here, and you will never be perfect, so you might as well stop trying to be. Make as much improvement as you can, keep at it, and know that, by practicing consistently, you'll continue to improve over time.

So, what can you do instead of correcting? Here are three good responses to your mother believing that she's only nine years old, none of which correct her:

**1. Acknowledge what she said.**

Respond to her incorrect info with, "Oh, isn't that interesting," or the like. You need to let her know only that you've heard and understood what she said. If she seems satisfied with that acknowledgement, go on with your day without correcting her.

**2. Then redirect to something related.**

Go with number 1 above, and then segue the conversation to something she can still discuss. "Oh, you're in 4th grade? I bet you know cursive writing. You've always had such nice handwriting. Was it hard to learn?" Without challenging her errors, let her talk about something she can still discuss. Big bonus points if you can segue into her telling a favorite story that, though you've heard it many times before, you listen and respond to with interest.

Don't redirect using the "Look! Haley's Comet!" approach. That is, don't redirect her attention by saying something totally unrelated. She might feel disrespected and not acknowledged and will need to repeat the behavior again and again until she finally gets her needs met, even if that need was only to be heard, accepted, or understood.

**3. Use a Therapeutic Lie.**

If her error puts her or others in real physical or emotional danger, you might need to use a Therapeutic Lie or Fiblet. For example, she doesn't recall she lost her driver's license due to traffic violations, but thinks she needs to drive to the office – even though she's many years retired – for an important meeting. Here's how this white lie to preserve her safety and dignity might sound: "Oh, I got a call a little while ago from your office. The meeting's been postponed. They'll let me know when it's been rescheduled. No need to go in today! Let's go make some lunch." (See section below for more on lies.)

Notice how each of these approaches requires you to know about your family member's life in order to customize the interaction in a way that really reaches their needs. You could try a more generic approach, but customizing is much more powerful and effective.

Many people aren't aware just how much arguing they do within their families – it can be staggering when they start really noticing. Some families argue more than others, but even families who think they're not correcting and convincing can discover they are when they start listening to themselves. Stopping corrections makes a *huge* difference, even more so through practice. It's really worth the effort.

> Patty and Jim (unrelated, they don't know each other) each have dementia, and each tends to not want to drink enough fluids throughout the day. Since elders are at extra risk for dehydration, their respective families really wanted to nip their refusal to drink in the bud.

> I learned from Jim's family that they'd set out three quarts of water in the morning, telling him he needs to finish all of them by bedtime. They'd fight with him all day, pouring him large glasses to drink from, which he constantly refused. "I can't drink that much," he'd say again and again. I suggested they switch to small glasses only, and to put away the quart jars out of his sight so he's not overwhelmed by the goal they've set for him. I also recommended they try some flavorings like a lemon or orange slice. This ended up being helpful, and the result was less arguing and healthier levels of hydration for Jim.
>
> Patty's family also wanted her to drink more. They offered a drink, which she'd refuse, saying she just had one a few minutes ago, which she hadn't. She wasn't lying. She just couldn't remember accurately. I suggested they not ask her if she wants a drink, but to bring her and themselves a glass and to drink with her. "Can you help me remember to drink? I'm just not getting enough fluids in the day," is what I suggest they say as they hand her a glass, so they can drink together. The social aspect of drinking together helps Patty do what she knows is polite: to take part along with someone who is asking for her help. She, too, improves her hydration.

While both these people had the same "symptom," we had to look at the specifics of each situation to understand how to construct an effective and individualized solution.

**Lies I Told My Family**

A few more words about Therapeutic Lies or Fiblets. For many of us, it's really hard to lie to our family members, especially to our elders. It goes against all the training we've had about being honest with those who mean the most to us. However, it's important to realize that when using a Therapeutic Lie, you're aligning yourself with what I think is a much Higher Truth: your mother can no longer can look after her own safety, and *you* must keep her physically and emotionally safe. Using a Fiblet sometimes can be the only way to achieve that in the moment. Therapeutic Lies don't necessarily work every single the time since many with dementia fluctuate in their awareness of our shared reality, but when Fiblets do work, they can work wonders.

> Gertrude had vascular dementia and sometimes, when she sundowned in the afternoon, she forgot that her father, to whom she was very close, was long gone. She'd be upset that she hadn't heard from him in a while. Where was he? What was he doing? Why hadn't he called?
>
> Her family would tell her he had died 40 years ago. Gertrude would be beside herself with grief at his passing, and enraged that no one had told her he was gone for so many years. This sad event went on about three times a week. I was called in to consult about her very high daily anxiety levels.
>
> Knowing that anyone who is told their beloved parent is dead, and who grieves and rages about it three times a week, is bound to be generally anxious, I recommended that they stop telling Gertrude that her father was dead, and use a Therapeutic Lie instead. The family then started to redirect her need to speak with her father, by partly using a Fiblet. "Oh, he's away on business. Let's try him on the phone later tonight." "You and your dad are so close… I know you miss him. I always marvel when I think about that time you two [insert a favorite story]…." This showed they both understood her wish to re-experience

> her closeness with him, and they helped her visit with him in memory. Checking back in a month's time, I found that Gertrude was consistently far calmer, and the thrice weekly grief-and-rage sessions were simply gone.

**Let Go of Being Right**

It's likely that, on a regular basis, you're more accurate in the things you say than your mom is accurate in what she says. That's not surprising. She has dementia, and her ability to think and remember is significantly impaired. But it's no longer important who's right or wrong. *You and mom are now living in a post-factual world.* When you're with her – and not correcting her errors, of course – you're going to have to let her be right, even when she isn't right factually. Part of not correcting is letting go of needing to be recognized as right when you are. Insisting you're right is only going to cause a fight that has no upside for either of you. You don't have to falsely agree with her, saying she's right. But you don't have to *disagree*, either. Let her have the last word. If you must respond, say something like, "Oh, that's interesting." Or say, "I love you." A non-contentious reply is going to pay both of you with dividends.

**Exercise Your Creativity and Intuition**

I hope it's clear by now that, to use the Four Dementia Keys well, you'll need to adapt and customize them to fit your mother's life history, preferences, habits, and interests. You'll also benefit from using your own intuition – that small, still voice that knows what to do without breaking things down into step-by-step, logical terms. With no one-size-fits-all solutions, you'll have to often "color outside the lines" to be most effective. This means experimentation and inevitable flops. Be aware that mistakes are part of the process. I make errors, too. But I also work at not beating myself up, instead focusing on learning from every mistake I make.

Personally, I love the experimentation and invention part of this work, though. Being creative is a joy in my life, and I hope by adapting the Four Dementia Keys to your family member's needs, you will also find pleasure in the creativity it requires.

I really can't stress enough that you'll make mistakes. Some errors you'll understand immediately, and others will leave you scratching your head for a long time, clueless. I believe that sometimes we can best learn by feeling free enough to try new things – *even if we mess up*. Don't be hard on yourself. Mistakes are bound to happen. So that you'll do better next time, just learn the lesson hidden within the error.

**Fast Summary – Chapter 6**

• Dementia Key #2 is about always customizing behavior handling.

• Dementia care is not one-size-fits all because every person with it is unique.

• Understanding the person's still-intact abilities is a powerful way to improve behavior handling.

• Exercising still-intact abilities helps preserve them longer and slow decline.

• The person's preferences still remain; respecting preferences is a powerful tool for care.

• Stop correcting, using facts to convince, explaining and arguing, immediately.

• To stop correcting takes time to learn, especially for families… keep at it!

• Instead, acknowledge, redirect and/or use a Therapeutic lie to preserve dignity and safety.

• Let go of having to be right.

• Calming your dementia crisis will take your creativity and intuition.

# 7. DEMENTIA KEY #3: COMMUNICATE WELL

The urge to communicate is deeply human. As many types of dementia progress, however, people lose more and more of their ability to communicate using language. This includes speaking, reading, writing, and understanding others when they speak. When confronted with a declining ability to use language that's complicated with intense short-term memory loss, caregivers need to use specific communication skills.

## Communication Is More Than Words

The most important concept in Dementia Key #3, one that I point to again and again, **is that communication is not just about what is said using language.** Communication includes *nonverbal* language, which we've talked about already in Chapter 5. We also talked about seeing behavior as carrying information we must decode in order find ways to prevent the behavior or reduce its recurrence #3 of the Four Dementia Keys, Communicate Well, adds some additional ways to more successfully interact with your mother.

## Communication's Rules of Engagement

You'll see with #4 of the Four Dementia Keys, Harmonize Stimulation (Chapter 8), that people with dementia experience many sensory processing changes due to their brain disease. I won't go into the specifics of that now, but I will make recommendations here for better communication that take these perceptual changes into account. When communicating with your mother with dementia:

- Use simple sentences

- Use helpful gestures and touch along with words to help her understand better

- Slow down when you speak; speak clearly

- Give her time to respond; it could be considerably more time than you need

- Don't hurry her when she's speaking

- Make sure you have and keep her attention – maintaining eye contact helps

- Speak with her on her physical level. Don't stand above her. It's bossy

- Suggest words when she's stuck only if she tolerates it (see below on Aphasia)

- Keep talking to your mother, even if she can no longer speak

- Don't rush in close to talk with her until you're sure her mood welcomes you near

- Approach her only from the front, not from the back or side

- Ask for or give information one little piece at a time

- After early dementia, use only closed – not open-ended – questions (see more on Decisions, below)

## Aphasia: The Pause That Frustrates

Aphasia means a person has trouble or a complete inability speaking the words they want to say. Folks with dementia often have some type or amount of aphasia, which makes communication difficult and frustrating. This frustration extends to both the person with dementia and those around them.

Remember in Chapter 5, I talked about how slowing down is a benefit that insanely busy folks can receive from caring for someone living with dementia? Aphasia is one of those symptoms that simply cannot be hurried. Hurrying not only fails to produce success, it's also upsetting to the person who's struggling to speak, which could make their speaking even slower. Take two frustrated people and put them together to communicate, and it doesn't take much to spark a fight. Stop fuming impatiently while you wait for your mother to find her words: she has aphasia and can't help it. She can tell that you're impatient from your body language, so just take a deep breath and be ready to wait peacefully for her to speak.

Some people with dementia do well with and appreciate it if you make suggestions for the word they're struggling to find. Others will show a flare of anger as soon as a missing word is suggested. And yet others will vary in their response depending upon the person offering the word, and the way it's offered. It's difficult to anticipate exactly what might happen if you help your mother find a lost word.

My suspicion is that her reaction has a lot to do with your body language, tone of voice, and whether or not she feels pressured to verbally complete her thought. This is another case where you should be creative and flexible, see how it goes, and adjust as needed. Or just stop it altogether if supplying a word makes your mom irritable.

I adore language, both spoken and written. I've been a professional writer and editor for decades. But I've had some aphasia since the head injury in 1992. It's much improved nowadays, but if you've heard me speak before an audience, I can get a little stuck at the end of the sentence – particularly with nouns. If it's a proper noun (like a name), regardless of where it is in a sentence, it's a struggle for me to speak it. I don't look or sound frustrated because I have developed a lot of practice with these pauses. I used to work hard to either come up with the problematic word or find a close substitute without taking longer than a very brief pause of maybe one "um" or "uh". It was exhausting work to be fluent before audiences, which I eventually did pull-off well. I don't believe anyone ever noticed my aphasia because I didn't call attention to it, and I was covering it up so well. But it was really exhausting to do!

More recently, I've given myself permission to relax and not fight my aphasia. Instead, I take a deep breath and let go of struggling when I do public presentations. (If you ever come to see me speak, you'll know this and certainly be patient with me. Your acceptance will help me be more relaxed and I'll stumble less…thank you). With friends and family, I often use less precision in my words, and I keep more accepting folks around me who help me relax so I retrieve better. I also feel free to just say "thingy" or "whatever" for any missing noun –I roll my eyes a bit and we have a giggle together. Just like other folks who have aged, I, too, care less about social conventions and rules that don't work well for me.

In a similar way, don't focus so much on your mother's frustrating aphasia, but look instead to being a patient listener. I suspect things will go more smoothly for both of you as a result, with fewer outbursts and possibly smoother speech.

Some folks with dementia have a workaround: they'll substitute another word, sometimes a nonsense word, for the one they can't say. Of course, this is a real conversation stopper. The temptation is to correct them, tell them that's not a word, and supply the one they're looking for, or give up because it's so bizarre. But Dementia Key #2 tells us not to correct the person with dementia. Instead, you can say, "Can you tell me more about it?' Alternately, ask "What's it used for?" or "Can you show me?" Questions such as these ask for more information without making the person with dementia "wrong." It's vital that you take these occasions to practice letting go of needing to be right with your mother. Without a word, this will help her let go of needing so much to be right with you, too.

## See Her as Whole

We often have a lot of preconceptions about people who are different from whatever we consider to be "normal." While we're conscious of only some of this bias, we can potentially become aware of more of it. Bias operates under the radar, coloring our perceptions and responses. Even healthcare professionals who work with elders have biases and direct stigma toward their patients and clients. It's an all too human situation. Like everyone, I'm involved in daily life-long training about how to view others clearly. I would be deluding myself if I said I were too evolved to have any biases, although I want to address and heal every one of them as soon as possible.

Your mom is acutely aware that there's plenty of stigma in society about having dementia. She knows this not just because she's on the receiving end of that stigma, but because before she had dementia, she, too, likely had some amount of bias toward others with cognitive impairment. The feeling of being singled out as abnormal is one that can make her hesitate to leave the house or see people. She's ashamed and might feel it's in some way her fault.

The toxicity of shame for being different is something very real to grapple with. Over the years, it's been freeing to have a variety of disabilities I've grown to be at home with, though I wouldn't have chosen to experience them. It's one of the most important gifts I received from being brain injured. I trust now if I'm relaxed about my cognitive (or other) disabilities, one or both of the following can happen, and they're both simply lovely:

1. What's wrong with me defines me less, both in my own eyes as well as those I'm interacting with;

2. I'm more relaxed and unselfconscious, so my energy is freed to be as capable as is possible for me.

By the way, this doesn't mean I don't ask for or accept help as needed. But I instead focus on what I can still do, which is what other people most often notice about me, too.

I suggest you spend time not talking about, noticing, or obsessing over, your mother's dementia disabilities. No, you don't need to hide the obvious or pretend that your mother doesn't have dementia. Just stop talking, thinking, and being hyper-focused, on how much she can no longer do. This is *especially* vital when you're with her because she can quickly pick up your "I can't believe she just did that" vibe and feel disrespected by it. We treat people differently when we focus only on their disabilities, and changing your focus and perspective can potentially iron out some of the wrinkles in your relationship with your mom.

Spend some time focusing on what she *can* still do, and the ways that she is still the same. Expressing happy recognition of her abilities is a form of positive reinforcement that will help both of you see one another as more whole.

At first, try this perceptual shift for just a few minutes, then fifteen, and then half an hour. Then work up to one or two hours. What might first happen is that you'll become more aware of your inner talk and how it's distorting your view of your mother. "I can't believe she doesn't remember how many children she has! And, oh! She's dropped something again! And now she can't get her words out. I'm going to scream. I am losing my mind," or a close version of these thoughts, might be what's constantly playing just beneath your awareness. All it takes to start moving away from this kind of incessant, automatic gloom and doom is to become aware of your inner talk.

With a bit of practice, you'll soon see your mother in a somewhat more whole and capable light. Since our expectations always color our outcomes, you'll end up with more of what you're looking for. You and your mom can find a new, more peaceful connection. If you like this feeling, keep practicing. Every day, make a point of focusing for a while on mom's still-intact abilities. It won't take any extra time. Simply do this as you go about your normal care tasks and interactions.

> Tasha was being cared for by a home care agency whose dementia program I created. Her family was asked to talk about her to help

> the caregiver learn about her personal history, likes, and dislikes. They expected this to be a tedious chore. Instead, they were surprised how much they enjoyed it. "I see her differently now," her son said. "I'm remembering her when she was more well and, actually, I see now that she's the same good soul inside." The others nodded in agreement. For them, it was like their familiar person had disappeared, but it was really their perspectives that disappeared her. Until they did this exercise, they had no idea that their perception had become so overcast.

**What Dressing Do You Want on That Word Salad?**

Some people with dementia keep speaking even after they no longer can construct an understandable sentence. It sounds just like language, but it's all jumbled up, so you can't get any meaning out of it. It's called "word salad" and it can make family caregivers *very* frustrated.

I've seen word salad in a variety of people with dementia. What I've noticed is that it's often conversational. You can hear the rhythm and cadences of conversation even though you can't understand the content. There's more going on than just an attempt at speech: the person is *expressing* themselves. If you stop listening for intelligible words, you can focus on your mother's body language and tone...and there's always a lot of meaning in body language and tone. Your mother is trying to connect with you, to simply shoot the breeze together or to share with you together. *Together* is the important word here: the folks I've seen tossing around word salad are looking for an exchange of words. They're looking for a *conversation*...a meeting of minds.

If a word salad is what you're hearing from your mom, just relax and let her word salad away. Mirror her emotions and you'll always fit in. If she's happy and relaxed, you should be too. Nod and say, "Uh-huh" and, "Oh, really?" like you'd do when listening to someone you completely understand. Laugh when she laughs. Be serious when she is.

I find these to be surprisingly satisfying conversations. So many verbal chats aren't necessarily deep in content – it's *connecting* that's important. It's as though the word salad strips down a superficial conversation to its very essence, *emotional connection.* It doesn't matter what you say in response. What's essential is *how* you act and look when your mother is speaking, and *how* you use words and gestures when it's your turn to speak.

For quite a while into their decline, people with Alzheimer's are very talented at this type of seemingly superficial conversation. This is how your mother deceived you and others into not noticing her dementia until it was impossible to hide it any longer. Families that live a distance away can be shocked to realize just how much impairment was going on, and for how long it was happening. "But we speak *every week*! How could I have missed what was going on?" is the most common reaction I hear from families just beginning to grapple with the presence of dementia in a family member.

The fact is that they're doing their best to cover their weaknesses (any or all resulting from their dementia) by hiding behind a strength (their chatting abilities). The truth can be even more concealed if they live with a spouse or partner. No, they're not conspiring to fool you. They're used to working together as a complementary team: when one is more forgetful, the other compensates. It was only after an in-person visit that you and your family realized things were far from going well in your mother's home. And this, too, is typical.

Don't be angry. It's a testament to what is still intact that makes misapprehension possible. Your mom wasn't taking pleasure in pulling a fast one on everyone. She was doing the best she could to cope. Some people who hide their growing cognitive impairment do so because of shame, or to prevent being placed in a facility, which is what they believe will be their family's response.

## Decisions, Decisions

There are many reasons why your mother has trouble making decisions, both large and small, since she developed dementia. Everything from deteriorating judgement, confusion, and not enough short-term memory to mull over choices, conspire to impede decision-making. However, your mom's going to be happier, more cooperative, and feel more in control, if you set up decisions for her in a way she can best navigate them.

#1 difficulty is that, in our industrialized Western culture, we have so many, many choices. Think about a visit to a sandwich shop. "Would you like white, wheat, rye, pita, a roll, bagel, multi-grain or gluten-free bread? Tomatoes, lettuce, onions, salsa, hot peppers, or pickles? Mayo, mustard, chutney, cranberry sauce or ketchup? Swiss, cheddar, American, brie or provolone cheese? Smoked, cured, spiced or boiled sliced turkey?" Even a so-called simple decision to have a turkey sandwich is too much for her to handle without help.

When caring for people in the middle stages of dementia, here are rules for setting up decisions so they can make them "on their own". As soon as you see that your mom struggles with making up her mind, remember the following:

• Never give more than two choices at a time ("Do you want white bread or rye?")

• As often as possible, show her the choices as well as articulate them

• Without rushing or pressuring her, give your mom literally as long as one minute to make even the simplest decision

A curious thing about giving someone with severe memory loss two choices is that they might no longer recall the first choice once they've heard the second. Their echoic memory (note the word "echo" here), the special type of very brief memory we have for things we hear, still operates. Echoic memory lasts only a few seconds, too short to capture that first choice. As a result, they'll typically choose the last choice they hear. (Now you can add echoic memory to another type of memory that is not destroyed early like short term memory is in Alzheimer's.)

> Before dementia, Cynthia used to love to put together her day's "look." She has a lot of clothing and accessories and she loved to look sharp. But now, she wore the same clothing for many days in a row before her daughter, Catherine, who was one of my students, could convince her to change. A year ago, Cynthia could change her clothes by herself, but she'd choose a light sleeveless cotton dress even if there was snow on the ground. She'd choose her Sunday best to go to the grocery store, and sweatpants for church on Sunday. Stripes were worn with checks, plaids with florals, and almost everything had food stains. Catherine was horrified and would make her mother change into different clothing. This was the subject of many, many fights between them that left them both angry and exhausted.
>
> There was more than one challenge going on here, several which you've probably identified. The biggest is that Cynthia had too many clothes to make the right choices. I taught Catherine how to stage decisions for her mother to make. The out-of-season clothing was removed to a different part of the house. Her very favorite things were kept in her bedroom closet and bureau, and everything else was placed out of sight in another room. Catherine learned to take out and offer just two outfits for her mother to choose from, and the outfits were always appropriate for the day's weather and activities. Being able to see and touch the clothing helped Cynthia decide what she wanted. As long as *she*

chooses what she wears, she's now more willing to change her clothing more often.

Catherine also came to realize that Cynthia's wearing the same clothes for a couple of days in a row wasn't worth having a big fight about. Ditto when Mom wouldn't change into night clothes to sleep. After a few weeks of this new routine, Cynthia's moods throughout the morning were clearly better because the morning didn't start out with a fight about getting dressed. Rather than focus on how her mother's clothing looked, Catherine paid attention to starting the day in a positive way. Her mother's dressing in clean clothes that were appropriate to the weather and occasion became the "new normal" that she and her mother liked.

## The Paranoid Things They Say To (and About) You

First, some terminology so we understand one another.

**Paranoia:** A paranoid person has excessive – often baseless – feelings of suspicion and persecution.

**Psychosis:** A psychotic person has had a loss of contact with reality. Their personality, thought processes, and behavior might be bizarre, and can include delusions and/or hallucinations.

**Delusion:** A delusion is one or more strong beliefs held by a person despite evidence to the contrary.

**Hallucination:** A person who is hallucinating is experiencing something through any of their senses that seems real, but is not.

The bizarre, suspicious, paranoid claims your mother makes can feel hurtful and frightening to you. These include claims that you've stolen her money, that someone's trying to poison her, that you didn't give her any dinner when she just got up from the table a few minutes ago, that someone's broken in and robbed her on a regular basis, etcetera. When faced with this, it's vital that you keep calm and remind yourself that she has a disease harming her brain. Please do not take her words personally, even if she means them to taken that way.

I know that it's not easy to keep calm when it seems insanity is on the loose, but I'm going to teach you a mental trick to give you something to focus on other than your feelings of being wounded or shocked. This shift in focus will also help you be a better caregiver because you'll come up with better solutions to calm your dementia behavior issues.

To me, more often than not, the "paranoid" statements your mom's throwing make *a ton* of sense. Here's why. Let's imagine that, a few years ago, you gave the family silver to your daughter. If you couldn't remember what happened five minutes ago, much less five years ago, and you couldn't find the silver in its normal place or anywhere else in the house, then it is *logical* to suspect that it's been stolen. But if your suspicions weren't viewed as memory loss, but as *paranoia*, you'd likely end up on antipsychotic medication.

These are drugs with the "black box" warning from the United States Food and Drug Administration (FDA) that begins: *"WARNING: Increased Mortality in Elderly Patients With Dementia-Related Psychosis,"* and continues for several more scary paragraphs in a similar fashion. Since antipsychotic medications don't treat memory loss, there's no reason to give them to someone who simply forgot that they gave away the family silver. In fact, in some cases, antipsychotics can worsen dementia symptoms and should be used only as a *last* resort. Unfortunately, despite the "black box" warnings and many professional initiatives to reduce the use of antipsychotics to treat dementia behaviors, these medications are still too often used as a *first* resort

You should understand wild accusations to be clues to how your mother perceives her world. Next time you hear suspicious accusations, first ask yourself what dementia symptoms are being expressed. As you learn more from the next chapters, you'll better understand your mom's "paranoia," "psychosis," "hallucinations," and "delusions" as the misunderstandings they might be.

While it is true that people with dementia can accurately be described as paranoid, deluded, psychotic and/or hallucinating, in my experience it is far more frequent than we fail to recognize the kinds of mistakes that people with dementia tend to make. Without a doubt, people with Lewy Body or Parkinson's dementia can have psychotic symptoms that are very real to them. However, I find it much less likely in people living with Alzheimer's, Frontotemporal or Vascular dementia.

**Fast Summary – Chapter 7**

- Dementia Key #3 is about communicating well.

- Communication is much more than words.

- For best results, always use the dementia communication "rules of engagement."

- Seeing a person with dementia as "whole" is a powerful tool for improving care.

- Keep talking to people with dementia, even if they no longer speak or don't make any sense when they do.

- Never give a person with dementia more than two choices to decide between.

- Use yes/no or either/or questions; do not use open-ended questions.

- Be careful labeling dementia behavior as "suspicious" or "paranoid" – it could just be a result of misunderstanding the world around them.

# 8. DEMENTIA KEY #4: OPTIMIZE STIMULATION

This chapter continues the theme of providing enough background info to help you use and customize Dementia Key #4. We'll now look at how your mother's immediate physical environment influences her ability to function, as well as her moods. We'll start with how dementia changes or distorts her senses, which are the gateway to her environment.

## The Senses in Dementia

Our senses are powerful tools that we use to navigate our world. They can become altered and vague – even absent – as dementia progresses. As a person living with dementia past its early stage, your mother, in her confusion and difficulty with language, generally can't report changes in how she senses her world. However, these changes can be observed in her behavior and ability to function. Because these changes actually have to do with your mother's brain, and not her eyes and ears, vision or hearing exams might not pick up on these problems.

## Boredom Can Drive A Person Out of Their Mind

As dementia progresses, although her senses may become dimmed, your mom still needs enough stimulation to make her feel good. Some of the mischief she gets into is because she's bored, and she's stimulating her senses without the filters of safety concerns and social acceptance. Unless you make a point of providing her with safe ways to stimulate her senses, she'll find ways of her own that you might not like in the least.

**Changes in Vision: Eye of the Storm?**

Some of the most profound sensory changes people experience as their brains decline due to dementia have to do with their vision. I don't mean that something about their eyes changes. How their brains process what their eyes see is what changes. Any eye diseases present such as glaucoma or cataracts may exist in addition to what we'll cover here.

When you see how profound these changes are, you'll quite literally have a much stronger sense of the world through your mother's eyes. These visual processing changes are so important to understand that you'll likely change many aspects of how you care for and interact with your mother in response. You'll also better understand why certain experiences you've had with her unfolded the way they did.

As people who have had a normal amount of visual ability all our lives, you and I are dependent on what we see in order to function at our best. Dementia seriously changes people's visual abilities in the following ways, each of which I'll discuss separately, adding information about how you might better help your mom deal with them. When people have dementia:

- They lose peripheral vision
- They need a lot more color contrast
- They need much brighter light
- They lose their depth perception
- They can lose half their visual field

Most families and healthcare professionals are never taught this information. I could rant for hours about this educational failure because, if you're *not* taught this information, you're unlikely to ever figure it out by yourself or grasp it intuitively. Therefore, if you've never had this information, you've been seriously handicapped in caring well for your loved one. No wonder caregiving has been difficult, yes?

**No "Side-Eye"**

Have you ever had a conversation with your loved one with dementia that went something like this?

> *Scene: [You place a cup of tea on the table in front of mother who is sitting on a chair at that table. A few minutes go by.]*
>
> Mother: I still am waiting for my cup of tea!
>
> You: But, I gave it to you already!
>
> Mother: No, you *did not!*
>
> You: What do you mean, I didn't?! I most certainly did. It's *right in front of you!*
>
> Mother: Well, it's not here!
>
> You: I can see it's there! I'm looking *right at it! [grabs hands full of your own hair and pulls them out of your head]*

If you've ever had a conversation like this, it was completely maddening because you probably didn't know that your mother with dementia could lose her peripheral vision. We depend greatly on our side, (also below and above eye level), also called peripheral vision, to get through our days, keeping us safe. "Catching something out of the corner of your eye" is a common English phrase, one that captures an experience we assume to be universal.

The normal human field of vision is about 40 percent peripheral, or side, vision. Occupational Therapist and dementia educator Teepa Snow says that by mid-Alzheimer's disease, people have only a 12-inch field of vision because their side vision disappears. Unless we know we're with a visually impaired person (or are so impaired ourselves), we're so used to having a normal range of vision that we assume everyone else does, too, and we interact with them as if this were always the case. But for yours and your mother's sakes, you must let go of this assumption. Imagine how differently you'll care for her now that you know how dementia compromises peripheral vision. Here are some relevant ideas that I've developed through my work with families experiencing dementia troubles:

- Place items you leave for your mother where she can see them as she walks around. Make sure to leave them at her eye level, around 4.5 to 5.5 feet, depending on her height.

- To keep from startling your mother, seat her at the dinner table so she can see the person who leaves or takes away her plate.

- Move her favorite chair to face front to something she likes to watch, such as the TV or a picture window.

- Always approach your family member from the front, not from the side or back.

> Hannah and her father had had a difficult relationship for many years now due to his drinking habits and tendency to rage at anyone nearby. Due to her father's dementia, she and her father had been living together for the past year. Hannah was most distraught by something new her father was doing, which was to often completely ignore her when she entered a room. Her father wouldn't glance her way, start a conversation, or even say "hello." She thought he was furious with her. Because he so easily exploded in rage, Hannah was afraid to say anything. She remained silent and left quickly the room, shedding a quiet tear.
>
> Hannah took a dementia course with me and learned about how many people with

> dementia come to lose their peripheral vision. "It's not that he's been ignoring me.... It's that he can't actually see me, and doesn't even know I'm there," she marveled. "He's also partly deaf, and he couldn't hear me, either. I had made up a reason why this was happening because I didn't understand his visual processing problem."

## High Contrast Living

Folks with dementia develop typically trouble identifying and understanding objects. I think this is in part because they have trouble clearly seeing objects are even *present* if the objects don't strongly contrast with their surroundings. In other words, a cup of black coffee in a black mug on a black tabletop could very well be invisible to your mother. When you realize that your mom can't understand or name what she can't see, you're less likely to be mystified by her behavior.

Color contrast comes into play throughout all activities and care tasks. This fact can be used to both conceal as well as to reveal. Yes, if you want to hide the cookies in plain sight, try using a cookie jar with very low or no contrast to its immediate environment. However, when it comes to concealing something hazardous or valuable, I wouldn't count on this technique working when your mother rummages through and explores things in the kitchen.

Here are two small stories that illustrate how using visual contrasts can make things easier for your mother to do with greater independence.

> Herby keeps frustrating and angering his wife, Milly, when he gets into bed at night. "You've done it again, Herby!! You're not under the top sheet, you're on top of it! What's the matter with you?" she grumbles. Once she learned about the need for high color contrast, she realized that Herby's dementia didn't let him see two same-colored bedsheets

as two separate ones. Now Milly makes the bed only with sheets that contrast strongly, and Herby can easily find his way between them at bedtime without Milly's help or objections.

Tony has been having new troubles in the bathroom, and his daughter, Maddy, who cares for him, was simply furious. "He's peeing in the waste paper basket next to the toilet!! I can't believe it! He's two inches from the toilet, and he can't make it except to go in the basket?!" Maddy wept and raged. We talked about what her bathroom looked like, and she described a space of light-yellow walls, tile, tub, toilet, toilet seat, and sink. The bathroom's waste basket, however, was made of dark-stained wicker. It was clear to me that Tony just couldn't find the toilet in that sea of light yellow. The dark basket was easy to find, round, and in almost the exactly right spot in the room for the purpose he used it.

As a result of our consultation, Maddy bought a light-yellow waste basket, and I gave her three choices for providing contrast in the toilet: cleaning tablets for the toilet tank that turn the water blue; a toilet seat in a color that highly contrasted with the toilet bowl and floor; or a colored light that's made specifically to illuminate the inside of the toilet bowl (honestly, these do exist and can be found online). Since her bathroom wasn't blue, Maddy chose to try the cleaning tablets, and they worked! But since Tony seemed a bit uncertain during his middle-of-the-night bathroom trips when the ambient light was low, Maddy later got a toilet light in purple. (Purple has the highest color contrast with yellow, being its complementary color.) Now, even in lower light, Tony doesn't get confused when he urinates. He even enjoys aiming at the purple-glowing water in the toilet.

**And Speaking of Low Light…**

As we age, we all need more light to see compared to the brightness we could manage with when younger. This is even more acutely the case with people with dementia. Your mother needs *two to three times brighter-than-normal-household light* to function at her best.

Imagine for a moment that you had one-third to one-half of the light you needed to read this book. How would you perform? Not well. And I'd bet you'd develop high frustration and eye strain pretty fast. If you give your mother the light level that's adequate for *you*, you're giving her one-third to one-half of the light she needs to function. Make sure she gets this extra light at least in bathroom, kitchen, hallways, and stairs, or your mom will literally be stumbling around in the dark.

**How Deep is the Ocean, How High Is the Sky?**

As dementia progresses, people can lose their depth perception. It takes too much work for your mother's ailing brain to correctly take the information from each eye and overlay those two views in a way that doesn't cause double vision. Without depth perception, your mother can't tell how far or close something is. *This is important for your mother's safety*. For example, she could try to sit on the edge of the bed, miscalculate how far it is, and fall to the floor.

She also can't tell if the thing on the floor is something she can pick up or is part of the floor covering's pattern. If you think she's trying to pick up things that aren't there, either you or her doctor might think she's hallucinating and she might end up being given antipsychotic medication for what's actually a mere trick of the senses. This is likely another part of why your mother can struggle to identify and understand objects. Things didn't used to look flat and, without a third dimension, they're harder to identify and understand.

Stereoscopic vision – seeing through two eyes – is a pretty neat trick, but it takes some brain power, which is exactly what fails in dementia. So, the brain, still more clever than we can imagine despite dementia, simply ignores the info from one eye. Everything

appears flat, or two-dimensional. But that's better than double or blurry vision, which can be both nauseating and difficult to navigate.

> Lucy was an "eloper," leaving her son's house whenever she could. "I'm going home!" she'd announce as she went for the door. Her family attended a class I held where they learned that people with dementia may avoid a black area on the floor because they interpret it as a hole in the ground.
>
> Lucy's family bought a wide black mat that was longer than the door was wide. Once they put the mat down, Lucy stopped trying to dart out the door. However, the mat also made her refuse to go through the doorway for a trip to the doctor because the black mat was at the threshold. Her family therefore learned to remove the rug when they wanted Lucy to leave the house. (This doesn't work for every person with dementia who elopes or wanders, but it's an easy enough solution to try out)

**Down to Only One Good Eye**

A further step in how folks with dementia lose even more vision, especially depth perception, is that their brains begin to ignore all the information received from one eye. Depth perception depends upon both eyes working together. Because it's difficult to coordinate information from both eyes, the information from one eye might be completely ignored by the person's brain. Loss of half the field of vision means that depth perception is eliminated. Also, all visual information about what's happening to one side of her is no longer on your mother's radar. It is as if she only had eye.

Therefore, in later stages of Alzheimer's, the 12-inch visual field decreases to a mere six inches. After decades of full vision, if your mother had only a 6-inch porthole to peer at the world around her, life would become a daily struggle for her and everyone around her. Since she has a very limited ability to retain what she learns, she won't be able to pick up the needed coping strategies widely used by other visually-impaired people. She might not be aware of or able to talk about her partial loss of vision, but be assured, it causes her to struggle with even simple things.

> Randy's daughter Charlene noticed that when he read, he would cover one eye. "It's just easier to read when I do that," he explained when Charlene asked him about it. The optometrist couldn't find that anything had changed in Randy's eyes since he'd been diagnosed with Alzheimer's two years ago. It was a mystery until Charlene studied behavior handling with me and learned how stereoscopic vision can be too hard for people with dementia. Now Charlene has one less thing to worry about. She understands that her father is doing exactly what he said he was doing. He's making it easier for him to read.

## Now Hear (Feel, Taste and Smell) This!

Let's look at dementia's impact on the other senses. Hearing is important for stimulating our brains. In fact, correcting hearing loss in a person with dementia can significantly slow down their decline due to their brain disease.

Hearing is a sense that can delight or over-excite those with dementia. The sounds of favorite music, for example, can be wonderful for stimulating the brain. But if it's too loud, too fast, or goes on for too long, then you might see dementia behaviors that tell you your mother is over-stimulated. Neighborhood noise, noise from machines and appliances, and too many people talking at once, can also be over-stimulation that quickly triggers dementia behaviors. Noise can be distracting, too, which doesn't help when trying to engage a person with dementia in a conversation they can understand. However, music *can* provide a positive distraction from care tasks your mother doesn't like.

> Giovanni loved to sing the popular songs from his youth, decades ago. Now with dementia, he hated having his nails cut, and he'd fight and scratch anyone who attempted to cut them. His son, Terry, and I talked over the situation, and when nail-trimming time next rolled around, Terry started singing Giovanni's favorite songs before he brought out the clippers, or even suggested trimming his nails. As Giovanni's lovely voice belted out "On the Sunny Side of the Street," Terry sang along, held his father's hands, and quickly clipped his nails. While Terry sang harmony to his dad's melody, his dad didn't notice what was going on. This not only got the job done without fuss, it gave Giovanni and Terry an opportunity to enjoy some wonderful time together.

When it comes to the tactile sense, your mother with dementia has a strong desire to touch and be touched by people, and by objects with interesting textures. In later dementia, your mom might focus even more intently on enjoying her sense of touch for pleasure and self-stimulation. We humans experience the most touch sensations in and around our mouths, hands, feet, and genitals. This is why research has shown that hand or foot massage can help relieve distress, agitation, and insomnia in people with dementia.

Like many folks with dementia, your mother has trouble keeping weight on. This can become a real health issue because it means she is continually under-nourished, which can lead to illness, even death. While there's a variety of reasons why this happens, one of the most common is that the senses of taste and smell have dimmed. Food just isn't as interesting. While this is true of people as they get older, in general, this loss of taste and smell is greater than what we see as a result of normal aging.

Or perhaps there was an earlier time when your mother gained an unhealthy amount of weight because she'd forget that she'd just eaten. She might have eaten lunch twice, for example, or dinner three times. And when she saw a cake in the kitchen, she'd take one slice, then another, until she'd eaten it all, not remembering how many slices she'd eaten. She just kept needing some interesting tasting food in her mouth.

### Sensory Stimulation: Too Much of a Good Thing, Or Not Enough?

We're all so different in what most pleases or upsets. The same is true among people living with dementia. When it comes to sensory stimulation, some people feel better when they have more of it. Others, better when they have less. Though people vary from day to day, and under different circumstances, a person who prefers a lot of stimulation cannot be fully transformed into a person who prefers a little stimulation, no matter how hard you try to transform them. Or *vice versa.*

People with dementia who need more sensory stimulation will be calmer and more upbeat when people visit them regularly, and have opportunities to talk, as well as listen to energetic music. They like strongly flavored food and drink, and more activity around them. People with dementia who prefer less stimulation will do better with more quiet, more time alone, listening to calming music, more bland food and drink, and less activity around them.

It's helpful if you accept how much stimulation your mother does or doesn't need. It's also helpful if you understand and respect her tendency to seek or avoid stimulation. Until you understand and respect your mother on her own terms, she'll likely either seek stimulation any way she can get it, or act out because she's overstimulated. With either response, the day probably won't go the way anyone wants it to. Take a deep breath and embrace your mom exactly as she is.

### Check the Environment

Check her environment for various environmental factors that can trigger dementia behaviors. Here are some to consider when trying to decode and understand their behavior, or when trying to prevent future dementia behaviors. Think about if something in the room is too:

- Hot or cold?
- Bright or dark?
- Noisy or quiet?
- Spicy or bland?
- Fast or slow?
- Colorful or muted?

Playing with adjustments to these and any other characteristics of her immediate environment can help you calm your dementia crisis, as well as teach you how to prevent or limit similar crises in the future.

> Nick was always a warm-blooded man who rarely was cold. His wife, Eleanor, was always chilly, even when wearing a light sweater in the summer. Now, however, they both had dementia. Because he felt too warm, Nick insisted that Eleanor not be given a sweater during the day, or extra blanket at night in bed. He'd also turn the thermostat setting down in the winter, chilling her even further. She was so cold, she would curl into a tight ball in bed, moaning and shivering.
>
> I brainstormed with Nick's and Eleanor's family, and they decided to disconnect the thermostat from the furnace and install a working thermostat in a less central location that would be difficult for Nick to figure out. He could change the settings all he wanted on the familiar, but now nonfunctional, thermostat, while the new one, out sight, did its job. Nick's family spoke with Eleanor's doctor about the problem at home, and the doctor wrote a prescription requiring that

Eleanor dress in one or two more layers of clothing than Nick did, and sleep with one or two more blankets than Nick required. For easy reference, the family taped a copy of the prescription on fridge, as well as on the wall in Nick and Eleanor's bedroom. The home care company made sure to enforce the prescription about blankets and sweaters. Relative peace, safety, and comfort were restored.

## Fast Summary – Chapter 8

- Dementia Key #4 is about harmonizing stimulation (not too much; not too little)

- Dementia changes how we process information from our senses

- There are vital changes in how a person with dementia perceives their world that we must know.

- Loss of peripheral vision in Alzheimer's and other dementias should change how we provide behavior handling.

- More light and color contrast, loss of depth perception and information from one eye also should change how we provide behavior handling.

- Hearing, taste, smell and touch also processing undergo changes that require changes in how care is provided.

- Too much or too little sensory stimulation are common triggers of dementia behaviors.

- Every individual's need for stimulation is varied and must be respected for best results.

- Environmental factors such as temperature, light, and noise can be common triggers for dementia behaviors if not customized for the person.

# 9. ADDRESSING THE PROBLEMS THAT DRIVE YOU NUTS

Hopefully by now you've been working with this book for a while and have been able to restore some sanity and peace to your life using The Four Dementia Keys. In this chapter, we'll discuss some challenges you might not have calmed yet. We'll look at three common – and very exhausting – situations so you can see how to apply the Four Dementia Keys to turn around your dementia trouble.

## Seeing The Four Dementia Keys at Work

While you might find many annoying behaviors fade away using just Dementia Key #1, it's likely there are others that are more complex and require some combination of some, or all, of the other Keys. I'll cover all Four Dementia Keys for each of these three care situations in a systematic way, which I want you to replicate in the areas where you're still struggling with your mother with dementia.

The format we'll use to troubleshoot dementia behaviors that don't easily diminish with just one or two interventions includes the following four steps:

1. Think about the behavior you're trying to turn around, about how it unfolds moment-by-moment.

2. Identify the pivot point(s) – those actions and moments where things start going wrong, or the moment before the

problem becomes apparent, and what tends to go wrong just before things take a turn for the worse.

3. Go through all Four Dementia Keys at the pivot points.

4. Decide what went wrong and what you'll do differently the next time this behavior arises.

**Example 1: Violence and Aggression**

Violence and aggression can include behaviors that make you or others feel unsafe, either by words or deeds. This includes hitting, biting, scratching, kicking, or threats to do so. Violence and aggression can go in either direction: *toward* the person with dementia or *from* them. Here, we'll look at them coming *from* the person with dementia. The next chapter, Chapter 10, will look at what to do if the violence and aggression are coming *toward* your mother from you or another family member.

Violence and aggression can often be found where out-of-date behavior handling methods are in use. Violence and aggression can happen when family caregivers are unprepared, or who have far too little help and support to handle the many challenges of caring for a member living with dementia.

By using The Four Dementia Keys consistently, much less violence and aggression should be taking place. If difficulties are calmed before they begin, they usually don't get so out of hand. However, sometimes a person with dementia can be unexpectedly triggered and overwhelmed by feelings that can escalate quickly to violence or aggression. Such is the following story of unexpected and uncharacteristic violence.

> Jasper was usually a peace-loving man who didn't fuss about much… until he developed dementia. His adult children were helping him since their mother (Jasper's wife) passed away in the spring. Jasper's children knew nothing about The Four Dementia Keys.

> Now it's winter, and he's become increasingly difficult to handle. He gets so terribly angry when they care for him – not all the time, but regularly enough that it's become something they're at a loss to handle. For example, yesterday at bedtime, Jasper's daughter, Winnie, tried to get her previously sweet, calm father out of the dirty pajamas he'd been wearing for eight days (and nights) and into something clean. "Dad, you smell bad! I can't believe how long you've worn those PJs! You have to change them!"
>
> It didn't take long before Jasper's protests and refusals escalated from pushing and shoving to hitting. Winnie now sports a black eye. Jasper was difficult to manage the rest of the evening until he finally got some sleep from 3-6 am. Winnie was up most of the night, too, and she and her siblings are exhausted, angry, ashamed of Jasper's behavior. They can't stand another night of this.

**Pivot Point(s).** Winnie demanded he change his clothing before bed. She said he smelled bad from eight-day-old pajamas.

**Dementia Key #1: Nurture Emotions.** Winnie got insistent about his hygiene right at bedtime, when they were both tired and when he needed to wind down peacefully for sleep. In order for Jasper to feel peaceful and safe, his bedtime requires the Dementia Two-Step. Instead, he felt traumatized and wound up. The problem with hygiene should have been handled more strategically at an earlier point in the day. Even though he quickly forgot the actual incident, Jasper was upset that things had gone badly with his daughter, and all night, except for too few hours of fitful sleep, he was agitated and unhappy. Avoiding these incidents at bed time will prevent his emotional upset when he's ready to sleep.

Jasper's pre-bedtime activities should have focused only on creating a calming atmosphere: gentle music, favorite picture books, sweet story sharing, and the like. But because Winnie was so upset and angry, she had avoided him and let him stew all night. Had she known she ought to at least try to calm her father, she would have felt calmer, too. They both would have slept better and felt closer. This mutually beneficial tranquility isn't always possible, but Winnie could have made progress toward achieving it if she'd been aware that her distress made Jasper upset and agitated.

**Dementia Key #2: Customize Care.** Because of Jasper's short-term memory loss, he thought he'd changed his pajamas earlier that evening. He couldn't keep track of many days he'd been wearing them, so Winnie's saying it was eight days seemed to him like an offensive exaggeration. Since Jasper was still an early riser, morning would have been a better time to get him out of his night clothes. When he was still working, he usually got dressed after he had breakfast. He'd take his second cup of coffee into the bedroom and sipped it as he got ready for his day. I recommended that Jasper's previous morning routine be maintained to help him feel better understood and more interested in changing his PJs.

Jasper also had a history of sundowning, which meant his confusion and other dementia symptoms would get worse starting sometimes as early as mid- or late-afternoon, sometimes continuing far into the night. This is yet another reason that night was the worst time to demand he do something outside his comfort zone.

**Dementia Key #3: Communicate Well.** Winnie corrected and argued with Jasper, which left him feeling insulted and upset. I suggested that, first thing the next morning, Winnie's brother, Dan, and *not* Winnie, who'd likely make Jasper tense, ought to make Jasper's breakfast and make no mention of the previous evening's debacle. I also suggested that, while Jasper ate, clothing that he enjoyed should be put out for him to put on after he'd finished breakfast.

I explained to Dan that when his father was done eating, he needed to say to him in calmest, most upbeat tone he could manage, "Hey, here's your second cup of coffee, Dad. I know you love to get dressed over your second cup. I'll carry it into the bedroom for you." And then, after guiding Jasper to his bedroom, he ought to say, "Oh, look! Your clothes are all laid out already. I'll never for that when I was a kid, Mom told me to take a sweater, and I wouldn't, and you talked me into taking one…what a blow up that was, but you made it work, Dad." I also told Dan that once Jasper had taken off his pajamas, they needed to be removed from his view immediately, or else he might want to put them back on again.

**Dementia Key #4: Optimize Stimulation.** That cup of coffee in the bedroom was a fragrant and tasty reminder to Jasper of what he'd done for decades when getting dressed in the morning. I advised Dan to hand his father the warm cup of coffee to increase his father's trust in him and make him more relaxed. I also suggested that Dan might add some calming music to his father's room. Additionally, I told Dan that he could help his father dress more easily by making sure that the colors and textures of his father's clothing contrasted strongly with their surroundings, like the bedspread they're laid out on. And by saying things such as, "Dad, that blue sweater is just perfect with your eyes. You always look great in that color. I love that Winnie knitted it for you," Dan could encourage his father to tune in to the enjoyment of the color and history of the sweater.

## Example 2: Midnight Freak-Out

Here we have an example of the many different ways people with dementia make errors because they no longer understand their environment or what their senses are telling them. When presented with their mixed-up interpretation, you'll have to untangle what they're talking about in order to prevent it from happening again, or to reduce the chance that it will.

> Before Pamela goes to bed at midnight, she wakes her mother, Joyce, so that she can empty her bladder. Pamela gets Joyce out of bed, turns on the lights in the hall and bathroom, and then lets Joyce go to the toilet on her own, since she is still able to do that task independently. About twice a week, after Joyce has finished on the toilet but is still in the bathroom, she starts yelling, "Hey! What

> are you doing?!! Get out of here!! GET OUT!!", and then emerges from the bathroom looking frightened. Every time this happens, Joyce explains to Pamela that an old woman had been looking at her through the bathroom window. Pamela always goes right into the bathroom, looks around, and announces, "There's no one in here. Everything is ok!" But Joyce remains unconvinced and needs a long time to calm down and get back to sleep. Pamela dreads this every night, wondering if this will be one of those nights. She just wants it to stop. She's going to call the doctor for some medication because her mother won't keep doing this.

**Pivot Point(s).** Something is happening in the bathroom that frightens Joyce.

**Dementia Key #1: Nurture Emotions.** Pamela's frustration is as reasonable as Joyce's need for reassurance. However, telling Joyce that there's nothing to fear makes Joyce feel that Pamela's ignoring what she's seen and amplifies Joyce's distress. By recognizing that her mother needs reassurance, Pamela could try saying something like, "Are you okay? I'm going to close the door to the bathroom, and I'll stay here with you here for a while. Then I'll be in the next room while you sleep. Just call if you need me. *You are safe with me, Mom.*" By staying in the moment and addressing what's immediately at hand, Pamela's approach is calming and supportive.

**Dementia Key #2: Customize Care.** What both Pamela and Joyce don't realize is that Joyce isn't seeing a woman in the window. *Joyce is catching a glimpse of herself in the bathroom mirror.* Joyce believes she's still a young woman and doesn't recognize that "old woman" as herself. This is a simple, common glitch in dementia perception that all too often ends up being diagnosed as a hallucination and medicated as such. But Joyce was seeing something real – *her own reflection*! It was no hallucination. Pamela deserves a lot of credit for maintaining Joyce's dignity by letting her go to the bathroom alone for as long as she's able to do so on her own. But had Pamela quietly observed her mother in the bathroom, Pamela might have caught on to the confusion that frightened Joyce.

Always be sure to inspect the site of upset for an explanation of what's triggering a seeming hallucination. Given that Joyce no longer recognizes herself, her understanding of what she saw was entirely logical. Because antipsychotic medications can be dangerous to the point of hastening death, they should be administered to people with dementia ***only as last resort.*** Fortunately, in Joyce's situation, a non-medical solution was at hand.

**Dementia Key #3: Communicate Well.** See "Dementia Key #1" above, which significantly overlaps here and would make the same points.

**Dementia Key #4: Optimize Stimulation.** Gentle music, calm talk, a hand or foot massage, or some other favorite calming activity, are needed in the moment when someone like Joyce is distressed. To address the problem in the long-term, Pamela could wait until the following day to take down the mirror, or turn it to face the wall if it's not built in like a medicine cabinet. If the mirror is built-in, Pamela could cover it with a picture or piece of fabric or a painting that Joyce particularly liked. If Pamela observes that during the day, Joyce can use the mirror without a problem, then Pamela could turn the mirror around or uncover it, or encourage her mother to use a hand mirror, or one that sits on a stand, that can be easily concealed at night.

**Example 3: Rummaging**

Rummaging is when a person with dementia looks through the drawers, cabinets, closets and other storage areas in the house, touching, holding, and maybe removing things that have been stored in them. As they go about rummaging, people with dementia may hide the things they find in unexpected places. Rummaging also often gives rise to two other trouble-causing difficulties. Your mother might get into dangerous, fragile, or valuable things that can be damaged or lost, or even harm her. And rummaging is a behavior that drives many families *crazy*.

> Sanford was in the latter half of Alzheimer's when he started rummaging in the dressers, cabinets, and closets throughout the house. He'd spend hours every day rummaging, which seemed to please and engage him quite a bit. His wife, Betty, was driven to distraction by this behavior. "All he does is make a mess! He pulls out everything from one place, and then moves on to another to do the same. I am so tired of cleaning up after him!" Betty lectured Sanford about stopping. She's told him to clean up after himself. She yelled at him. She went so far as to turn the dressers around so Sanford couldn't get at the drawers… but neither could she. Sanford broke crystal glasses and misplaced important papers. Betty found a credit card bill shoved down the kitchen garbage disposal. While she appreciated this likely unintentional attempt at humor, Sanford's rummaging antics rattle her and kept her on edge.

**Pivot Point(s):** No specific triggers for Sanford, but Betty's the one with the outbursts here because Sanford isn't always safe rummaging, and too many things have been broken or lost. Betty is constantly cleaning up; she is losing her temper.

**Dementia Key #1: Nurture Emotions.** Yelling, threatening, bargaining, and the like weren't only failures as solutions, they left both Betty and Sanford deeply unhappy. Sanford is beyond engaging with Betty in the sort of rational discussion that would end in his agreeing, and remembering, to never rummage again. In place of pointless discussions, the Dementia Two-Step is what's needed.

**Dementia Key #2: Customize Care.** I advised Betty that she and her grown children box up, remove, or lock up, her valuables, and anything that's fragile or dangerous. Sanford would never need fancy stemware, so crystal glasses could be squirreled away without causing him upset. Stuff that's necessary for day-to-day life but potentially problematic might be stored in cabinets or closets outfitted with baby-proof latches or locks. And anything irreplaceable, or otherwise of extraordinary monetary or personal value, ought to go to a friend or relative for safekeeping.

But Sanford needs to be allowed *at least* one dresser, closet, or cabinet, filled with harmless non-fragile things he can happily and safely rummage through. Better yet, these rummage-worthy objects should be easy to pick up and clean when Sanford is done with them. To prevent mail from finding its way into the garbage disposal or being otherwise misplaced, Betty can ask her local post office to hold mail for pick up, or forward it to a family member or friend willing to sort through it with an eye to making sure that Betty's and Sanford's bills are paid, and otherwise keeping their finances in order. Betty and I also needed to talk about her letting go of the urge to immediately clean up after Sanford's rummaging. Since it's a chore that's easily undone, perhaps Betty could make a separate peace with tidying up only once, instead of several times, each day?

**Dementia Key #3: Communicate Well.** Once all had been made ready for him, Betty needed to bring Sanford to his own rummaging spot. I suggested that she open the drawers or doors and show him what's inside. To encourage Sanford to rummage responsibly, I told Betty to touch and look at the objects with him, and then leave him with his pastime once he was engaged in that activity. At this point in Sanford's life, showing him was better than telling him.

**Dementia Key #4: Optimize Stimulation.** Sanford uses rummaging for his daily enjoyment and should be permitted to continue, but in a safer fashion. He simply doesn't know what else to do with himself since he can no longer independently start, continue, or finish an activity. Rummaging is freestyle, which suits his diminished abilities just fine. If Betty truly wants to stop the rummaging, she would need to provide him with another equally pleasurable activity he's able to do on his own, or with ongoing observation and help. All things being equal, it's better to allow him to continue to rummage under more controlled conditions.

Sanford especially loves to touch and look at the woolen gloves, hats, and mittens in the front hall cabinet. And *this* was a perfect solution. Let him rummage through the woolens cabinet however much he wants. Since these are things that can be quickly and easily be put away, Sanford can have a blast rummaging through them again and again. If he ever seems at loose ends, Betty could bring him to the woolens cabinet and start rummaging with him. He'd soon be engrossed and able to continue on his own. And once Sanford's favorite rummaging spot has been established, Betty needn't search the house for him like she had to do before.

I could of course add many other examples of dementia behaviors that need a combination of all Four Dementia Keys. But the overall troubleshooting of behaviors that don't easily recede includes these steps:

1. Think about the behavior you're trying to turn around, about how it unfolds moment-by-moment.

2. Identify the pivot points, or those actions and moments where things start going wrong, or the moment before the problem becomes apparent, and what tends to wrong just before things take a turn for the worse.

3. Go through all Four Dementia Keys at each of the pivot points.

4. Decide what went wrong and what you'll do differently next time this behavior arises.

## Fast Summary – Chapter 9

• Use all Four Dementia Keys to prevent or address dementia behaviors

• Spend a moment considering each Dementia Key individually to find triggers for the behavior as well as ideas of how to intervene successfully.

• Identify "pivot points" where the behavior began, reflecting on what was going on just before and during those moments to find triggers for dementia behaviors.

• Come up with a plan to use to prevent a recurrence.

• This book's Toolkit contains helpful documents that can be downloaded from www.FromCrisisToCalm.com.

# 10. CARING FOR THE CAREGIVER

***Don't skip this chapter!*** **Really, you need what's in here.**

I say this because, too often, while family caregivers will help others until they drop off their feet, they'll be just as quick to skip a chapter called "Caring for the Caregiver." Any dementia caregiver who won't do self-care eventually falls apart, which quickly leads to their family member becoming much harder to deal with. When a caregiver falls apart, what tends to happen is that the person with dementia becomes too difficult to be cared for at home and must be placed in a facility.

*But this so often can be avoided!* In fact, that's exactly what we're here for, you and me, *to help you and everyone in your sphere get through your mother's dementia without breaking*. To restore some sanity and peace to your life, you must also turn some attention to yourself as an at-risk family caregiver.

## You're the Lynchpin, Driving the Care and Household

If you are focusing only on your family member's needs and don't take adequate care of yourself, you'll eventually have a crisis, or series of crises. If you have a crisis, your loved one with dementia is going to have one, too, only bigger and much worse.

Research from the United Kingdom shows us that, if family dementia caregivers become depressed, there's a much higher risk of their family member being institutionalized. This research also reveals that this can be avoided with a combination of sufficient support, education, and self-care. Looking at your own well-being is unavoidable if you're going to calm your dementia crisis and keep it that way.

Of course, since you're learning the material in this book, behavior handling *should* get easier. While behavior handling will never be easy as pie, it doesn't have to be as terribly difficult as many folks experience it. That's a bit of welcome relief!

**"If I Have to Take Care of Myself, Too, I'll Scream!"**

I hear you now, muttering under your breath, "Good grief, don't give me more that I have to do!" Now, isn't that exactly the problem with self-care? It's just another @&#$ thing to do, and your plate is already so full. *I so get it!*

But what are your alternatives? If enough self-care doesn't get onto your to-do list regularly, then the eventual consequences might be far, far worse. You might develop **Caregiver Burnout**… or you might already be there. Let's take a look at CGB (Caregiver Burnout) so you understand how much is at stake.

CGB happens when caregivers have reached a state of such exhaustion that it becomes difficult to carry on with a caring attitude. Body, mind, and spirit are so worn out that the caregiver is negative where they used to be upbeat, uncaring where they used to have concern. They're squeezed dry and perhaps unable to experience positive emotions. CGB is caregiver stress on steroids – an afternoon off or a massage isn't enough to provide relief even in the short-term.

It can take up to a couple of years for full recovery from CGB, even after the person stops caregiving. There are aspects of CGB that are just like Post Traumatic Stress Disorder (PTSD), which also takes time and attention to put into remission, much less resolve. If you're not there yet, then it's vital you prevent Caregiver Burnout. We'll talk more about how in Chapter 11 when we look at some additional resources you might need to help you.

## Caregiver Burnout...It Can't Be *Really* Serious, Can It??

Actually, CGB can be and is truly serious. There are real and potentially long-lasting health consequences with CGB. For example, for stressed elder family caregivers, their mortality rate is 63 percent higher than for adults who aren't providing family care. This is mortality from *all* causes. I find this troubling, and while I'm not trying to frighten you, I do hope that you'll take this information to heart, too.

Research also says that, due to the stress of the job, family caregivers can die up to *ten years earlier* than people who aren't family caregivers. What's more, family caregivers also experience more obesity, a diminished immune response, slower healing of wounds, and a greater vulnerability to diabetes and other serious illnesses.

There are 168 hours in a week. But did you know that only nine hours of weekly family caregiving can *double* a woman's risk of heart disease? How many hours a week are you caring for your mother with dementia? I bet it's more than nine.

Here's the last statistic I'm going to throw at you about this. While I really do enjoy irony, in this case it's just too insulting to cause me anything but distress. It turns out that family caregivers are at *higher* risk for cognitive impairment. This means that, if you wear yourself out taking care of your family member with dementia, you'll end up with a greater risk of needing the same type of help sometime in the future.

## How Do You Know if You Have Caregiver Burnout?

Let's take a closer look at CGB. Please note that there are many symptoms that overlap with depression and PTSD. Distinguishing between the conditions isn't our task here and, actually, it doesn't matter a lot. If you are depressed, experiencing PTSD or are burnt-out (or all three), you need this information all the more. CGB symptoms include:

- Withdrawal from friends, family, and activities you previously enjoyed

• Preoccupying concern about the future

• Feeling blue, irritable, hopeless, and helpless

• Changes in appetite, weight, or both

• Changes in sleep patterns

• Getting sick more often

• Feelings of wanting to hurt yourself or the person for whom you're caring

• Emotional and/or physical exhaustion

• Excessive use of alcohol and/or sleep medications

• Irritability, moodiness, negative responses and behavior

• Difficulty concentrating

• Preoccupying wish to run away or escape from your life

**Call your local emergency number or the National Suicide Prevention Lifeline at 1-800-273-8255 if you feel you might hurt yourself or someone else.**

## Self-Care Is a *Must*, Not a Luxury

Does this list sound like you as a result of caring for your mother with dementia, while trying to keep up with everything else? If so, I strongly urge you to consistently adopt some combination of several of the following recommendations:

• Accept that you have limits

• Be realistic about what you can do within those limits

• Work with a coach specializing in dementia and behavior handling skills

• Accept you need help from others, and then get that help

• Learn and regularly use tools for self-care

• Join a caregiver support group, especially if it has a skill-building component

• Confide in trusted friends, family, or helping professionals

• Set aside time for yourself

• Find ways to laugh, keeping your sense of humor

• Accept that finding it hard to provide care is normal, not a character flaw

• Self-care is absolutely necessary – get enough exercise, rest, sleep, and healthy food, etc.

• Keep up with your own medical care, including when you're sick and as prevention

Find information about additional resources to help you prevent or recover from Caregiver Burnout in Chapter 11.

**Let Go or Be Dragged**

The letting go that a family caregiver needs to do in order to adequately address their own mental and physical health requires some significant adjustments. If you've read up to this point, and worked diligently with The Four Dementia Keys, you've already got a good head start on the letting go process.

Already, you've learned that letting go is part of good behavior handling. You've learned to let go of having to correct your mother. You've learned to let go of your personal baggage. You're checking it at the door so you can improve your behavior handling results. You're letting go of hurrying through care tasks. You're letting go of having to be right. You're letting go of out-of-date care methods. And, most difficult of all for a lot of people, you're letting go of many preconceived notions and expectations about dementia and behavior handling.

Until now, we've not talked about letting go of parts the old role you used to have with your mother, and your mother's role with you. Your relationship has changed in many ways because she needs you in new ways. Your feelings for and about her might have changed, too. I want you to know that this is quite normal, even expected. But it's important to let go of those old roles, because you won't need them with her again.

It's also normal that it's journey to let go of the idea that the present should look like the past, that your mother should be how she used to be. Life brings many changes, some which we like, others we don't care for one bit. At the very heart of the Universe is change – it's woven into the very fabric of our lives. We all change. We all age. It is in the correct order of things.

Every person who lets go of fighting what we cannot change becomes mentally and emotionally stronger. You can be, too.

If what I just said in the paragraphs above has made you feel uncomfortable, pissed you off, or made you say "No, it's not normal or expected!", then acceptance is part of the leading edge of your personal growth and learning.

This very worthy life work is beyond what we can cover in this book. With such types of growth, some more customized professional support can be very helpful. A psychotherapist, counselor, or personal coach could be very helpful to you if you're having trouble accepting and flowing with the facts of your current life.

**How Do You Know When It's Time to Get Help with Care?**

I also hope you're letting go of any role that looks or feels like dementia slavery. Don't go this alone with your mother once she's at mid-dementia or later (earlier if you or she have other significant physical or mental health issues, or you're providing care for more than one person). At that point, it's more than time for you to look for some substantial help from others. Otherwise, you can start help as early as you'd like to, though few people start care too early… too late is much more customary.

Here are the topics I'd talk over with you about getting help if I were your behavior coach, and we were doing a coaching session. First, what aspects of handling dementia behavior are your most and least favorite? What brings you the most satisfaction? What sucks your soul dry? I'd recommend you keep doing your most favorites, and get someone to help with the ones you hate. Your satisfaction and wellbeing are worthwhile including in the equation.

Next, I'd point out that whatever is special between the two of you should be preserved. Is it the special way you help your mother put on her clothes or change her adult briefs? Or is it special that you're her daughter, in a lifelong relationship?

I'm going to vote for the latter. A home care worker can help with care tasks and household chores, *but only you can be one of your mother's nearest and dearest.* If all your patience, time, and energy are tied up with her hands-on care and running the household, you won't have anything left to be a family member. With the right help, you can instead focus on putting your energy back into continuing the relationship you two had. It won't be exactly the same, but it'll likely be a more satisfying than focusing only on the chores involved in daily care.

In our session, we'd also discuss how daily caregiving can bring a new intimacy to your relationship that you've perhaps never had before, and whether you both want that or not. Between spouses, this is often more mutually welcomed than between a parent and child. Your and your mother's feelings are going to be quite individual and varied from others whom I've asked these questions, and there are numerous "right" answers. Only thinking and feeling your way through these questions will bring you to the answers that are right for you and your mom. A coach can help with that feeling-out process.

After that, we'd talk about the many types of tasks needed to keep your life and household going. For example, you love and adore the hands-on care role the very most and you handle it well. Then you need some other kind of help – cleaning, yard work, errands, banking, bookkeeping, paying the bills, shopping, home maintenance, providing funding for help, and so on.

These tasks should be distributed among family, friends, organizations that provide support for disability and/or the aged, and among those you hire for the work. Some of the above roles don't require in-person attention and can be taken on by distant friends and family. Hands-on daily caregiving is something I find quickly exhausting, but I can coordinate care and caregivers until the cows come home without undue fatigue. That's not a better choice, just one that's specific to me. Your choices should fit you best.

It truly takes a village to care for someone with dementia through the course of the disease. If you're a lone cowboy (or girl) who's not getting sufficient help and support from others, you're endangering your wellbeing and that of your loved one. More on how to select appropriate care in Chapter 11.

**Respite May be the Best Bit**

Respite (RES-pit) care allows family caregivers to get a break to rest and refresh themselves. Done at consistent intervals, respite care helps family caregivers like you stay fresh enough to continue providing high quality care.

Caregiver Burnout results in care so poor that your mother gets sent to live in a facility. Respite care does the opposite. It allows your mother to live well at home longer. Research shows that respite care can:

- Improve family functioning
- Increase life satisfaction for caregivers
- Increase caregiver capacity to cope with stress
- Improve attitudes toward the family member with dementia

- Reduce hospitalization rates for family caregivers and their loved ones

- Create a more effective and compassionate family caregiver

- Reduce stress in the home

- Increase marital stability

Research shows that it takes *as few as four hours a week of respite care to make a real difference*, though many families can greatly benefit from longer hour of care.

**Guilty, Guilty, Guilty**

Caring for your mother can be an occasion to feel oceans of guilt. There are many aspects of caregiver guilt, and I'm first going to address the one I most regularly coach families about. As I said in Chapter #1, many family members feel guilty that they weren't already using the most effective and up-to-date behavior handling methods.

As much as it's impossible to use methods of care you've never learned, guilt isn't rational. Guilt can truly take on a life of its own, leaving you feeling remorse and anxiety so broad and generalized that it's about everything and yet nothing. This is a reminder about what I asked you to align yourself with in Chapter #1. If you can't shake feeling guilty for not having been born knowing this book's contents (I wasn't either, by the way), please spend time daily reflecting on this affirmation:

**I have been doing my very best.**

**Now that I know better, I will do better.**

**New skills take time to develop until they become automatic.**

**I forgive myself, and let go of any guilt or anger I have.**

**I forgive myself for not being a perfect caregiver.**

And speaking of free-floating guilt: you can end up tortured by feeling that you somehow could have done something differently that might have prevented, slowed down or prevented your mother's dementia. This is a highly corrosive and destructive belief because we really don't know what *exactly* causes, slows down or arrests dementia. Research shows us many possible paths to prevention and delay, but we're not there yet with any real certainty, much less a well-constructed, evidence-based program.

Therefore, this type of guilt is actually as unreasonable and irrational as many of your mom's dementia-generated ideas. Give it perhaps a good five years from the publication of this book before you insist on feeling guilty about this. Hopefully, we'll know more about how to slow or arrest dementia by then, and you'll know exactly what to feel guilty about. In other words, please let go of this worry for now.

Feeling so guilty about committing to regular self-care that you just won't do it creates an intolerable level of self-neglect. If you're awash in guilt, or any other type of unmanageable feelings, please assess whether you might have Caregiver Burnout and seek help and support immediately. While guilt isn't yet a well-recognized symptom of CGB, it really should be on the list.

**Begin to Turn Around Burnout with a Simple Exercise**

Human beings engage in inner self-talk all the time. Despite its repetitive nature, we might not be aware of it. Detecting hidden self-talk is amazingly helpful, but it's not always easy to dredge it up to consciousness.

Burnt-out family caregivers often end up focusing on what has gone wrong, what is lost in their relationship with their loved one, how impossible it is to change what's happening in their dementia crisis, and on all sorts of other deeply unpleasant negative thoughts and emotions. This is a symptom of their burnout, if not its actual cause. (Researchers have proposed that CGB is an inability to experience positive emotions.) If you're conscious of your own self-talk, check out if that inner voice is giving you an all-cloudy view of your local weather.

Given how difficult it can be to hear your own self-talk, I think it's much easier to simply *change* it. The following very simple daily exercise was tested for two to three weeks on burnt-out medical residents – it's a version of Sheryl Sandberg's approach she wrote about in *Lean In*. It proved to be potent stuff for the medical residents, starting to turn around their physiological signs of burnout in just a few days.

Even better: when testing them a year after the initial study period, half of the medical students remained more resilient, resilience being a preventive for burnout. They also had significantly less depression and relationship conflict, and a better work-life balance. Only a few minutes a day for two or three weeks a year of doing this exercise is needed to achieve these powerful results.

And now you, O burnt-out caregiver, can do this exercise, too. It's called The Three Good Things, and here's how it works. At bedtime, spend a few minutes jotting down three things that were positive in your day. They don't have to be on the scale of "World peace began today." They could be enjoying a segment on the news, or that you were in a good mood all morning, which helped brighten your mother's mood, too. It's up to you what you choose.

Then assign one of these ten positive emotions to each of the three good things you wrote down: joy, gratitude, serenity, interest, hope, pride, amusement, inspiration, awe, and love. For best results, do this nightly for two to three weeks. Please keep in mind, however, that the researchers reported the Three Good Things process was a little addicting. Go ahead and document your Three Good Things for as long as you enjoy and benefit from the process.

What happens as a result of the Three Things exercise is that your inner voice stops noticing only the negative parts of your life and starts adding balance by highlighting positives, too. That black cloud that's been following you around starts to lift, and your psyche begins to notice the rays of sunshine between the clouds. It works… *start it tonight.* What have you got to lose? To help you with this process, find a Three Things worksheet in the From Crisis To Calm Toolkit. Download it from www.FromCrisisToCalm.com.

## "I Put My Mother in a Facility...I Feel So Guilty"

The last aspect of guilt I'll cover is the guilt you might feel if you place your mother in a facility such as assisted living or memory care. This is a highly-varied situation that needs its own section here.

People with dementia can live the rest of their lives at home, without a facility. Typically, this results in a higher quality of life for the person with dementia. At home, Dementia Key #2 can be in full force: Customize Care. A facility can't be expected to customize care for every resident to the level that can be done at home. This includes all care tasks, activities and environmental factors.

The highest necessity for a facility placement comes when it's a far better choice for the central people involved: the person with dementia and/or their primary family caregiver(s). Most commonly, I see the following three situations come up in the families I support. See if any of them might potentially fit the situation with your mother:

- **Not enough care is available at home.** This is when there's no way to provide adequate care and supervision at home, or the cost of doing so is unaffordable compared to a facility. The care needs of someone with dementia become greater with time, becoming 24/7 at late-stage or earlier. If her needs exceed your family's caregiving capabilities and resources, I believe a placement could be warranted.

- **Home is unsafe and cannot be made safe.** Some people with dementia hoard or haven't maintained their home for years, thereby creating an unsafe environment. Perhaps the design of their home has too many stairs or other impractical features. Alternately, they might be living with someone who abuses or neglects them, abuses drugs or alcohol, or otherwise makes the environment unsafe. In that case, I believe a placement is warranted.

- **The person with dementia is impossible – and maybe always has been.** The very best behavior handling methods – all my most magical moves rolled into a ball and dusted with fairy dust – cannot lessen the full impact of someone with a really, really difficult personality...including one that predated dementia's arrival. Lifelong intense dishonesty, self-centeredness, manipulation, intense control dramas or neediness, rage, greed, and meanness, leave some elders understandably without anyone willing to care for them. If this is what is between you and your mother, this might have made you miserable at times – or *all* of the time – until you left (*if* you did leave).

Dementia will likely make her only more intensely difficult in the exact same ways she always was. Trauma from your relationship is going to push yours and your mother's emotional buttons again and again. Yes, it would be an opportunity to work on your own issues because they'd be triggered a lot. But you aren't required to take on your mother's care as your personal-growth project. Unless you're really willing and able to do that inner work, provide behavior handling, plus her care, I believe a placement may be warranted.

**Are Others in Your Family Being a Pain Your Butt?**

Do your siblings get in your face, telling you how to care for your mother with dementia? Or are they just absent, leaving you to "do it all" by yourself? Or somehow both? I've noticed an unspoken rule at work in many families I've coached: the person who is furthest away (geographically and/or emotionally), and who has the least idea about what day-to-day life is like for the primary caregiver, is the clearest and most emphatic about what the primary caregiver is doing wrong and how to fix it. My suspicion is that what's going on that the more-distant person is feeling guilty because they're not closer. If this is what's going on in your family, I recommend a couple of things to help them become more a part of the solution:

**Establish that someone needs to be the primary caregiver and/or care manager.** The "primary caregiver" is the person who provides most of the care, and is likely emotionally and physically closest. As the primary caregiver, you often overlap roles with a care and/or household manager who coordinates medical and household tasks. When you make decisions about finances and medical needs at the point your mother with dementia cannot make them herself, it is because you hold their power of attorney/ conservatorship, health care proxy, and guardian, roles. That's a lot of roles for one person. Sometimes these roles can be better distributed across the family, such as someone not involved day to day takes on the financial roles.

However, the roles are filled, the primary caregiver needs the authority to carry out their day-to-day work without constant criticism and obstruction. The rest of the family should not be second guessing the small stuff unless it points to larger issues, such as neglect, abuse, Caregiver Burnout, or some type of really inappropriate care methods.

If the primary caregiver says, "We have to work on Dad's night time wandering," others focusing on changing different behaviors are going to dilute and confuse the effort, so they need to knock it off. There is only one primary caregiver, and your family usually knows who that already is, whether they admit it or not.

As the primary caregiver, you should have day-to-day authority to provide care and direct others who help with that care. Other roles need their own authority, too. Stop second guessing each other to death over little things. Big issues need a family meeting, perhaps with a mediator. This doesn't mean you don't consult with your siblings or let them know what's happening – typically, you should do both as the situation warrants.

**Give everyone specific jobs to do.** Even if they're physically distant, your family members (or very good friends) can handle banking, bookkeeping and insurance issues. A professional in these areas can also be hired – make sure they are insured and bonded. If your siblings are emotionally distant but local, they don't have to do hands-on care, but they *can* shop, cook, do yard work, take the car for repairs, and so on.

**Don't choose people to work together who dislike one another a lot.** Does one of your siblings tend to pour gasoline on your mother's emotional fires? There's no reason to say that all tasks family members help with need to be directly for the person with dementia. Helping you as the primary caregiver with your life (shopping, errands, financial support, and so on) can free you up to give better quality care to your mother. No use trying to mix folks who can't get along with daily dementia care – it won't work.

**Use a family mediator.** It's important to not let old, unsettled family baggage get tangled up with your mother's behavior handling needs. Some relatives will continue to seek resolution – or retaliation – for old family grievances with others, including the person with dementia. They don't likely reveal their grudges through a forthright discussion, but rather express them in the way they act around and treat others in the family. They may not be particularly aware they're doing this, but it's a need that's not appropriate – or maybe even *impossible* – to fulfill as dementia progresses.

Yet, they can and do persist in wanting to engage. How to refocus their energy? Though there's usually a peacekeeper in the family who wants to fill that mediator role, that's not necessarily the best approach in many cases. I think there needs to be a skilled, impartial person mediating serious family disagreements around caregiving. This could be a professional mediator, a family therapist, or a clergy person who is adept at mediation.

**Take a deep breath and let go.** Some primary caregivers can be inappropriately controlling, and don't work well with others who want to help out. If you're hearing from multiple people that you're keeping everyone out of the loop or being too bossy, then you should consider if you're holding on too tight to your role. When a primary caregiver steps into the relationship between the person with dementia and other family members, this is often going too far – unless others are putting the person with dementia into physical or emotional danger.

Even if everyone in your family is using The Four Dementia Keys, their creativity will lead them to interact in ways different from what you're doing. But is that necessarily a *bad* thing? You each have unique relationships so we have to expect the Four Dementia Keys will be deployed somewhat differently by each person who uses them. This is why dementia caregiving is an art. You need to let go of trying to control everything and everyone in the environment because that can create conflict. This kind of tension will set off your mother's dementia behavior, as it would with most people with dementia.

That said, there should be consistency in some areas regardless of who helps with care: the daily schedule, and use of The Four Dementia Keys or some other standardized system of modern behavior handling.

**Get help if a family member is endangering your loved one.** This can be abuse or neglect of various types: financial, emotional, physical, sexual, etcetera. I don't care what the politics in the family is like, a vulnerable person unable to care for themselves is being harmed. Call your local police department or adult protective services (more on the latter in Chapter 11) and make a report. The call can usually be made anonymously. Any helping professional or police officer is required to report elder abuse – you can reach out for help to one of these mandated reporters.

## Are You Scared You, Too, Have Dementia?

Seeing your mother's dementia up close, have you developed a terror that you, too, are developing some? You know how many things you're forgetting, that you're struggling with day-to-day chores, and wonder if it's genetic and you have dementia, too? Let's tease apart the issues here, because there are several overlapping.

The first issue is if dementia can be inherited. Some forms can be, but this is not even close to the majority of cases. Usually, it's not a sure thing, even if you have a heritable type of dementia in your genetic make-up, that you'll definitely have dementia. There's one genetic strain found in a very large family in the country of Colombia that's really certain to cause young-onset dementia. If you're not a member of this Colombian family, then your inheritance isn't certain.

The second issue is that with (non-Colombian) inherited dementias, it's not clear what triggers the gene to become the disease. In fact, the cause of most dementias is not truly known. Lifestyle factors seem to be in play. But which aspects of lifestyle are the most important to address? Good question. Researchers are working on their favorites, and someday we might know for sure. Then you'll know exactly what to do to reduce your risks. Right now, there are programs for brain health which might be helpful; give one a try.

The last issue tangled up here is that the changes we see in people's memories as they age normally are quite different from the memory changes we see in dementia, which is not normal aging. Some middle age (or older) family caregivers get in a literal panic that they have dementia because they don't know the several distinct differences between normal aging and dementia. You can find these differences spelled out in the online From Crisis To Calm Toolkit, which you can download free at www.FromCrisisToCalm.com.

Please look at this information – there's no use being terrified of something that doesn't apply to you. You don't have time or energy for unnecessary concerns. If you remain worried after looking through the criteria in the Toolkit, then please see your family physician. If you're under 65, don't let a doctor dismiss your concerns about your brain's health because you're "too young" – as the Baby Boomers age, we're seeing a spike in young-onset dementia (that's before 65 years old). Anyone who pooh-poohs your concerns isn't keeping up with current developments in their field.

Most family caregivers who worry they have dementia are actually so exhausted and stressed that their brains can't function correctly. No one's brain could work well under similar conditions! Your stress and fatigue is a huge reason your brain isn't working well. Please turn to the Caregiver Burnout section at the start of this chapter to think about CGB for yourself.

## Fast Summary – Chapter 10

• You MUST do regular self-care – this cannot be escaped without great consequences.

• Caregiver depression can lead to the person with dementia being institutionalized.

• Caregiver Burnout (CGB) is a serious condition that can lead to illness and early death.

• CGB can be prevented, minimized and treated successfully.

• Use the Three Things exercise to start to relieve CGB quickly.

• Respite care can be a powerful help; use it regularly, even as few as 4 hours/week can help!

• People with dementia can live at home the rest of their lives as long as it can be made safe enough, and there is enough care.

• There are solid reasons to have a some with dementia live in a facility.

• There are many roles family and friends can play to help; each person in their role should be allowed to do their job without micromanaging.

• Contact law enforcement or Adult Protective Services if someone is harming a person with dementia.

# 11. EXTENDED SOLUTIONS TO YOUR DEMETIA CRISIS

Though there are scores of pages in this book, there is *so much more* we could talk about when it comes to cutting-edge, evidence-based behavior handling that I haven't covered yet. Recall the purpose of this book is to get you out of your family's dementia crisis, because living in a state of emergency is too unsustainable and painful for another moment.

You might have been living in a state of *"just make it stop happening,"* where all you want is some relief for a while. Perhaps you've forgotten what life was like without an ongoing behavior handling crisis woven through every moment, both awake and asleep. This happens with many family dementia caregivers; they need support re-entering and re-tooling life outside of crisis.

**What's Possible After the Crisis is Over?**

I hope by now you're not living in an acute, ongoing dementia crisis, which means you're ready to start having your life back again, even if that life means resting and recovering! And so is your mother with dementia. There's a *huge* difference between not being continually distressed and having a good quality life. It's entirely possible to care for a loved one with dementia where life is mostly good for everyone involved.

> Rick had young-onset Alzheimer's, which meant he and his wife, Catrina, were still of working age. As his ability to function declined over time, Catrina's time and energy to work, care for him, and run the household – never mind do anything for herself – just wasn't enough. She tried home care, but Rick shut down emotionally as soon as the caregiver arrived. They tried several different caregivers, but he refused to speak with them, allow them to help him, or even stay in the room with him. This was very upsetting because Rick had always been such a warm people-person.
>
> Months went by and Catrina was so exhausted and stressed she again tried home care. This time, however, she chose an agency whose caregivers I had trained in dementia care. The first moment the front door opened, Rick smiled and then hugged the caregiver, whom he'd never met before. They sat together to chat and they cooked together much of the day, a favorite pastime of Rick's. They laughed and joked and had a marvelous, relaxed time together. The next day, a different caregiver from the same agency came, and together they had another wonderful day. "I've got my husband back! It's been so long since I felt like he was with me, but he's here now," Catrina said, with tears streaming down her cheeks. "I'm so grateful...I had no idea this was possible." She also now had time to see friends, keep her hand in activities she's interested in, get medical care for herself, and some much-needed alone time.

"What happened here?", you might be asking yourself. It's clear to me that since two different caregivers on consecutive days yielded marvelous results, that *excellent behavior handling skills can be learned.* Each caregiver knew after they rang the front doorbell, that they must Leave Their Baggage at The Door and Spread Some Sweetness starting the moment the front door swung open. Successful home care like this really changed Rick and Catrina's lives for the better; it might do the same for you and your family.

Results this good don't happen with everyone, but Rick and Catrina aren't just some oddball exception – there are many, *many* cases I've been involved with where transformation happened in the person's functioning, personality, relationships and moods. People with dementia, when they're able to communicate, are clear: *they are "still here."* When we learn how make them happy, to reconnect them to our shared world, how to communicate with and stimulate them properly, we can feel they're "still here" much more often. The full presence of their personality isn't usually on display 24/7, and it would be unusual to easily see someone be fully present in late dementia because of their communication and movement disabilities. But they're all "still here," where or not we can see it. None of us are obviously ourselves when we're high upset or checked out.

But if you can help your mother be calm and present to enjoy life, there is a world of wonderful things that become possible. Meaningful activities, warm relationships, sharing of fun and laughter, sweet closeness...even moments of joy. Your mother won't be just like she was before dementia, but you can still enjoy her as she is. There are many more improvements that could be made once your dementia behavior issues have been calmed.

**Assemble Your Dementia Team**

Dementia is a long-lasting disease. Therefore, care isn't a sprint, but a marathon. You'll need a team to help you through the different stages and the many ups and downs you'll experience with your mother. Throughout this book, I've promised you more information about several resources; now we're finally here. These include services you need now or will soon need, plus others I hope you'll *never* need...they're here just in case.

**Home care** is something I love, and I've worked in this field for years. I really want people like your mother to live at home as long as possible, including through their very last breath if conditions allow. Home care involves what's called "custodial care," which includes companionship, personal care, and home making. Sometimes, the home care is particularly enlightened and includes behavior handling training and meaningful activities; this is the best choice if you can find it.

However, the quality of dementia care offered in home care across the industry is highly varied. Every agency *says* they provide dementia care, but do they actually provide substantial training for their caregivers? Do their supervisors know how to coach and mentor behavior handling skills, especially as your mother declines and these skills need retooling? Or is dementia care just something they list on their brochure, without much actual substance? Most often in my experience, the latter is the case: a claim with little backing. Surprisingly, I've also seen that where's there's been caregiver training, the skills taught are never coached or mentored beyond that. This is almost a guarantee that within a month, the caregiver will stop using what they've learned. What a shame that is, because there aren't enough good dementia caregivers to go around.

Also important: will the agency teach, mentor and coach these skills so *you* can be a good care partner with them? Excellent behavior handling means that *everyone* in your mother's world needs to use the same care methods, so you and your family *must* be brought into the learning. The agency you use might not exactly know The Four Dementia Keys, but they should be using *similar* care practices, because regardless of name, these are the very ones needed.

I'm hesitant to recommend freelance caregivers, which means hiring a caregiver who don't work through an agency. There are so many vital issues I think families aren't aware of but should be. For example: families don't realize that when using long hours of care, they might have in reality become an *employer* (not a client) to that caregiver, which requires tax withholding, workman's compensation, overtime pay, and unemployment insurance that you would have to set up and fund. You'd have to follow all state and federal laws involving employees and withholding of taxes and insurance.

If you don't do this and a freelancer is injured on the job, they could sue you for a substantial amount. How likely do you think it is that your homeowner's insurance would cover the injury of your *employee* working in the home because you don't have workman's comp set up for them? I don't think they would. And if you get caught not withholding all those taxes, you'll might not only have to pay what's owed in arrears, potentially for years of work. There could also be substantial penalties.

If you don't pay overtime properly, you might again be liable for the missed pay and penalties. Is the freelance caregiver also bonded and insured? An agency should be, but individuals often are not. Not to mention: what happens when the caregiver wakes up with the flu and it's your day for their help? Who covers for them? Make sure *all* these issues are clarified to your satisfaction before you hire a freelance caregiver. Sometimes the lower hourly fee a freelancer charges can become *painfully expensive* in the end.

Consult with your lawyer, accountant, and insurance agent to know how these issues apply in your location.

**Adult Day Services** can be another wonderful opportunity for people with dementia. The best ones I know provide transportation, activities, a quiet room, meals and snacks, and an on-site nurse for up to 5 days a week. However, adult day services also must apply the same type of really good dementia care training for staff and families as home care services. Ask if you can observe them work for a couple of hours; you'll likely get some sense if the staff is really connecting and effectively working with their clients with dementia.

**Respite care** can be found through home care (see above), or it can be found at some facilities for a week or two, where the person with dementia lives there during that time. The same thing goes for respite as it does for home care and adult day: they must have specialized training in dementia, and actually use those skills consistently when providing care.

### "What Else Do I Need to Learn?"

Everyone needs a sense of meaning and purpose, including people living with dementia. In fact, they're often in desperate need to re-discover what kind of meaning and purpose they still have, especially as their dementia continues to unfold. A lack of these is a big trigger for dementia behaviors.

#1 of the answers to this need is going to be meaningful activities adapted to your mother's disabilities and abilities, including activities that reflect her long-standing personal interests and preferences. The skills needed to do this well are specialized – though I bet you can guess some of how this works just from your study so far of The Four Dementia Keys. These specifics are beyond the scope of this book.

Also outside the scope of this book is how to tweak The Four Dementia Keys throughout the whole course of your family member's dementia. Your family member starts out in early stage with pretty decent capabilities. You'll see many of these diminish as her brain disease progresses. How do you best adapt care to anticipate and respond to her newest losses? How can you help her when she starts having swallowing issues? How can you figure out what she needs once she can't speak? How do you get her to bathe without a battle? Each of these need to be learned and customized for your situation.

### Consider a Geriatric Care Manager (or Aging Life Care Professional)

Geriatric Care Managers have recently changed their name to Aging Life Care Professionals (ALCP). They are typically nurses or social workers who holistically work with older adults and others with health challenges in the following areas: health and disability, housing, financial, family issues, local resources, client and family advocacy, legal issues and crisis intervention.

I've found ALCPs can be very helpful when the person receiving care is medically or psychologically complex, or both. I often recommend an ALCP when family members can't agree on their loved one's needs, or when they don't agree there even *is* a need. A full needs assessment by such impartial professional can be very helpful, and it comes from outside your family's inner politics. They'll also create care plans, coordinate with other professionals, and advocate for the clients' needs. They can be local eyes and ears where family is far away. They also know local resources to guide you to. There are so many roles they can fill as needed, that I think the best of all worlds is using an ALCP practice that includes both a nurse and a social worker who work together.

ALPC's receive a substantial referral fee for placing a client in a facility. I consider this a conflict of interest – how can you know they are suggesting the right thing for the client when thousands of dollars are the result of such a placement? Because of this, I think this financial relationship needs to be put out on the table for families like yours. While they night not tell you how much they receive from each facility, the fact of compensation should not be an unnamed elephant in the room. Ask about it if you work with an ALPC.

Not all ALCPs know and use cutting edge dementia care methods. You'll need to interview them about their training in dementia care. Also, observe anyone you're thinking to hire with your mother to see how they interact. Find a certified ALCP through the Aging Life Care Association online at www.aginglifecare.org.

**Your Area Agency on Aging (AAA)**

If you're in the United States, the entire country is covered by geographic Area Agencies on Aging (AAA). Find your local AAA by searching on www.ElderCare.gov. There are over 600 AAAs in the US serving people aged 60 and older. They're funded by federal, state, or local monies – or some combination of all three. Each one provides a different selection of services, most commonly these are:

• Nutrition counseling

• Meals on Wheels or other feeding programs

• Long-term care ombudsmen

- Insurance counseling, including Medicare
- Elder abuse and neglect reporting
- Caregiving support and training
- Transportation help
- Referral and information services
- Connection to local resources

There is typically an income-based fee schedule for any paid services provided.

**If There's Abuse or Neglect**

Your dementia behavior challenges can include someone abusing or neglecting your mother with dementia, including another family member, friend, health care worker, or care facility – or maybe it's you. This abuse or neglect can include finances, sexual assault, medication, physical battering, threats, emotional abuse, withholding of nutrition, and so on. If you're in the United States, to talk with Adult Protective Services (APS) about elder abuse or neglect, call your local AAA. (see above section for more information, or go to www.ElderCare.gov).

The National Center on Elder Abuse has many online resources that might be helpful to you – look on their page under Resources for your state APS Office at www.ncea.acl.gov. **Call the police at 911 (or your local emergency number) now if someone you know with dementia is in immediate, life-threatening danger.**

**Medical Cannabis and Dementia**

There is both experimental and anecdotal evidence that using cannabis has real potential as a treatment for dementia, to address dementia symptoms and behaviors, and even to possibly prevent dementia or slow its progression. If your crisis is not yielding to the other recommendations in this book, or if you're looking to create the best quality of life for your loved one as well as the rest of the family, I recommend considering its addition to the daily regimen.

This is not meant to simply "drug out" your mother with dementia. Used at the right levels and in the right way, cannabis can help address the many symptoms of dementia without laying her out cold. It's areas of benefit include addressing loss of weight, pain, anxiety, depression, aggression, combativeness, withdrawal, sleeplessness, and distress. For people with dementia, cannabis has also been known to increase their ability to communicate verbally, be appropriately engaged with tasks and other people, to show more affection, and a whole host of other benefits that vary by individual. Your family member with dementia might be more like "themselves," too. This is an acceptable use of cannabis in many states that legalize this medicine for medical – or even recreational – purposes.

Cannabis can also be a godsend for family caregiver use. The stress, insomnia, anxiety, and depression you experience are also worthy of this beneficial medicine. I suspect it can also prevent or address caregiver burnout. Like with your mother, this isn't about drugging you out, but as a beneficial treatment so you can better cope with what's on your plate.

Good choices for delivery are vaping, or using topicals, tinctures, or edibles. Smoking cannabis is not recommended for health and safety reasons. It's best to find a medical practitioner who is trained and specializes on cannabis medicine to help guide you.

Using cannabis effectively as a medical treatment can require more knowledge and experience than you gleaned through your recreational use in high school and college; there are even three certification organizations through which licensed medical doctors and nurse practitioners can become board-certified in cannabis medicine. Search online by including your state (if in the United States and your state allows medical cannabis) and the term "cannabis practitioner". I recommend the same even in areas where recreational cannabis is legal so you can get the best guidance possible.

## Hospice Can Be a Real Godsend

I truly adore a good hospice experience, and I can't encourage you enough to bring one into your life as early as hospice will agree to provide services for your mother – typically six months prior to her anticipated passing. Please, oh, please, *oh please: do not wait until the person's very last days to start hospice* – you'd be missing so much potential goodness. With hospice, you'll get a big, experienced professional team to help meet your *whole* family's needs.

Hospice can help in so many areas I can't list them all here, including patient care, symptom management, and emotional and spiritual support. Hospice staff would address whatever issues are most important to your mother's needs and wishes. Hospice would focus on improving her and your whole family's quality of life. They do marvelous follow-up for months after your mother has passed, offering grief counseling and supportive referrals when additional help is needed. *Do not pass up these and many other benefits that last for six months or longer!*

They not only offer these precious, much-needed services, but research shows that hospice care on average *extends* life, not ends it sooner. Currently in the US, hospice care can be renewed after six months if need be. Or, your mother might do so well on hospice that she'll "graduate" for a period of time. She can always be re-admitted when it's appropriate. Hospice for the full period she qualifies can help calm and prevent dementia behavior crises, and help you *all* live better. I couldn't wish for better care for you, your mother, and the rest of the family.

While hospice is often feared as a sign of "giving up hope," in my mind it's *exactly the opposite*: s*tarting hospice shows that hope continues.* Hospice can help the end of life be full of meaning for everyone involved. Providing the best chance for peace of mind and comfort in the last six months of life is surely an act of great love and hope, not one of abandonment and hopelessness. Know that you have a right to interview and select the hospice that appeals to you most.

## Your Personal Dementia Behavior Coach

Because every person with dementia is such an individual, and there are so many changes your mother will go through, you might want to have additional help and support beyond this book. If you'd like to work with a behavior coach on any or all these topics (plus others), there are several available here on planet Earth, including myself. I offer training, group coaching and one-on-one consultations to families just like yours.

Go to FromCrisisToCalm.com to see or sign up for what's available for ongoing help and support. I would be honored to continue to help your family out of their crisis, and to then support you all to a higher quality life through every stage of your mother's dementia.

**Fast Summary – Chapter 11**

• The intention of this book is to help you end your family's dementia behavior crisis, and get back to living.

• There is still much to learn and that can be improved after your family's dementia behavior crisis is over.

• The right care methods can help people with dementia be more "themselves."

• The quality of dementia care varies enormously in home care, adult day programs, respite care, and so on; be selective.

• You will need the right team of professionals to help through the entire course of dementia.

• As disabilities increase over the years, you must continually adapt care to match the person's "new normal".

• Your personal behavior coach can help meet your and your family's needs as dementia progresses over time.

• Go to www.FromCrisisToCalm.com to find help and support for your family's behavior trouble.

• This book's Toolkit can be downloaded at www.FromCrisisToCalm.com.

# 12. CONCLUSION

Everyone family's journey into – and out of – dementia behavior issues is unique...as individual as each family. What these all have in common is a need to much better understand dementia, behavior handling and how to adapt this information to the situation.

Your mother can't teach you how to care for her in a new way, the way that all family dementia caregivers need to learn. She never learned how, and now likely couldn't muster the words she'd need to explain.

Research shows us that this type of non-intuitive (even *counter-intuitive*) care needs to be taught, and then mentored and coached for a period of time afterwards. The focus here has been to calm your family's dementia crisis by teaching – and mentoring – you in some of the skills no one ever told you you'd need to care for your mother with dementia. This included:

- Just enough information about dementia to strengthen your behavior handling

- The Four Dementia Keys:

    - #1: Nurture Emotions

    - #2: Customize Care

    - #3: Communicate Well

- #4: Harmonize Stimulation

- Addressing problems that make you the most nuts
- Caring for the caregiver
- Expanded solutions to your dementia behavior troubles

I'm having trouble ending this book, because I know that during the years she has dementia, your mother will need you in new and different ways as her abilities continue to decrease. I also know you'll need ongoing help and support, especially if you want to never get into the depths of crisis you've been in. I trust The Four Dementia Keys completely; practiced well, they'll see you through many future challenges. Just keep adapting them as your mother's needs change.

I hope if you need further help and support that you'll look for it on this book's website. The online, downloadable Toolkit for this book includes several documents to help you reach and maintain the even keel this book supports you to reach. Find it at www.FromCrisisToCalm.com. Do look around that site for public presentations, courses and other offerings that may appear there.

I am available, too, for consultations. If you're interested in further coaching, I offer a brief consultation to make sure that I am the right person to help you and your family. Email me at help@DecodingDementia.com to set up an appointment, or schedule one via the website.

I hope the time we've spent together here has managed to calm your family's dementia crisis, and that your family is enjoying each other more. I also hope that peace, warm relations, fun, laughter, closeness and pleasure prevails. Dementia care isn't easy, but it doesn't have to be as hard as it's been for you and family. With the right approach, caring for your mother could end up being a satisfying and deeply meaningful time for both of you. I'd hate for you to miss out on what could be such an important, life enriching experience!

**Fast Summary – Chapter 12**

- Remember you are doing your best, and that you will not be perfect at behavior handling.

- You must use behavior handling best practices to prevent crises and give the best quality of life possible for you and your family, including the person with dementia.

- You may need additional support and instruction over time; a behavior handling coach can help, including Dr. Bier.

- This book's Toolkit contains helpful documents that can be downloaded from www.FromCrisisToCalm.com.

# ABOUT THE AUTHOR

Deborah Bier, PhD's work has successfully trained thousands of family and professional caregivers in dementia care best practices.

Her credentials are unique in that she experienced years of cognitive impairment herself following a brain injury in an auto accident. She has been a psychotherapist and homecare agency director of care for over 25 years.

She is considered a gifted health educator, able to synthesize complex health topics into easy-to-understand language and down-to-earth actions.

Dr. Bier holds a PhD in therapeutic counseling, a certificate in gerontology, and is a Certified Alzheimer's Educator, a Certified Dementia Care Partner and a Certified Dementia Practitioner.

Website: www.DecodingDementia.com
Email: help@DecodingDementia.com

# THANK YOU

I'm honored you've chosen this book to help you through your family's dementia behavior issues. I'm grateful to readers, clients, and students like you who have allowed me into your lives at such a stressful time. I'm so impressed with the beautiful work you've done as you learned how to create calm, peace, warm relationships, and sweet times as together we've calmed your dementia crisis

As a thank you, I offer you the online Tool Kit that's a companion to this book. Go to www.FromCrisisToCalm.com and click on the Tool Kit tab to take you to this free resource.

I wish you, your mother, and your family much joy, closeness, and ease in the time you have together. I've seen the possibility for so many sweet, loving times returned to families who thought those days were long gone because of dementia. But they didn't know or practice what you've learned in this book. I hope you'll see a re-emergence of the person you've known as these decades.

Now, go practice your Four Dementia Keys. Brush up on your Dementia Two-Step. You've got this!

# ACKNOWLEDGEMENTS

We all stand on the shoulders of those who came before us, and I have been a student of some of the best who preceded me. Dr. Paul Raia, Teepa Snow, Joanne Koenig Coste and others who earlier laid a path of hope through the dark woods of dementia care and behavior handling.

To the many people with dementia and their families whom I've served, to the students I've trained, and the professional caregivers I've supported: it has been my greatest honor to be on this journey with you. Thank you for showing me what you most needed, and how you best could receive it. I also thank the many home care agency owners and staff who gave me years of affirmation and feedback on my dementia care and behavior handling coaching mojo.

I thank The Difference Press, Angela Lauria and her staff, who helped me see the potential for this book and for giving me a much wider view of my work. I am grateful to Abe Opincar for kicking me in the tush when I didn't yet realize I had a particularly powerful way of presenting health information that was much needed in our world. I am just delighted to have worked with Brooklyn Billmaier who created the cover and prepared the final manuscript for the e-book.

And finally, to my husband, Rich, who patiently awaited my distracted attention as I wrote this book, my eternal thanks for the million ways you make life more worthwhile.

Made in the USA
San Bernardino,
CA